QUALITY and GMP AUDITING

Clear and Simple

James L. Vesper

CRC Press
Taylor & Francis Group
Boca Raton London New York

CRC Press is an imprint of the
Taylor & Francis Group, an **informa** business

CRC Press
Taylor & Francis Group
6000 Broken Sound Parkway NW, Suite 300
Boca Raton, FL 33487-2742

First issued in paperback 2019

© 2010 by Taylor & Francis Group, LLC
CRC Press is an imprint of Taylor & Francis Group, an Informa business

No claim to original U.S. Government works

ISBN-13: 978-1-57491-055-1 (hbk)
ISBN-13: 978-0-367-40090-3 (pbk)

This book contains information obtained from authentic and highly regarded sources. While all reasonable efforts have been made to publish reliable data and information, neither the author[s] nor the publisher can accept any legal responsibility or liability for any errors or omissions that may be made. The publishers wish to make clear that any views or opinions expressed in this book by individual editors, authors or contributors are personal to them and do not necessarily reflect the views/opinions of the publishers. The information or guidance contained in this book is intended for use by medical, scientific or health-care professionals and is provided strictly as a supplement to the medical or other professional's own judgement, their knowledge of the patient's medical history, relevant manufacturer's instructions and the appropriate best practice guidelines. Because of the rapid advances in medical science, any information or advice on dosages, procedures or diagnoses should be independently verified. The reader is strongly urged to consult the relevant national drug formulary and the drug companies' and device or material manufacturers' printed instructions, and their websites, before administering or utilizing any of the drugs, devices or materials mentioned in this book. This book does not indicate whether a particular treatment is appropriate or suitable for a particular individual. Ultimately it is the sole responsibility of the medical professional to make his or her own professional judgements, so as to advise and treat patients appropriately. The authors and publishers have also attempted to trace the copyright holders of all material reproduced in this publication and apologize to copyright holders if permission to publish in this form has not been obtained. If any copyright material has not been acknowledged please write and let us know so we may rectify in any future reprint.

A CIP record for this book is available from the British Library.

Library of Congress Cataloging-in-Publication Data available on application

Visit the Taylor & Francis Web site at
http://www.taylorandfrancis.com

and the CRC Press Web site at
http://www.crcpress.com

CONTENTS

Dedication   **viii**
Preface   **ix**
Acknowledgments   **xiii**

1. What Is a GMP or Quality Audit?   **1**
Goals   1
Why Conduct GMP Audits?   6
Terms Related to GMP Audits   8
What an Audit Is Not   10
Conclusion   11

2. The Big Picture   **13**
Goals   13
The Audit as a Quality System Component   13
Five Phases of the Audit Process   14
Management Support   17
Monitoring Improvement   17
Outcomes of the Audit   17
Conclusion   18

**3. Regulatory Requirements and Expectations
for GMP Audits**   **19**
Goals   19

Requirements vs. Expectations   20
Conclusion   30

4. Your Auditing Policy and SOP   **31**
Goals   31
Policies and Procedures   32
Regulatory Inspection of GMP Auditing Policies
    and SOPs   33
Certification of an Audit   35
Regulatory Inspection of Vendor Audits   36
Conclusion   38

5. Types of Audits   **39**
Goals   39
Types of Audits   39
Making Use of Various Types of Audits   40
Conclusion   47

6. Audit Approaches   **49**
Goals   49
What Is an Audit Approach?   49
Audit Approaches: Quality System   52
Audit Approaches: Documentation System   54
Audit Approaches: Deviations   56
Audit Approaches: Validation and Change
    Control   57
Audit Approaches: Stability Testing and
    Monitoring   60
Audit Approaches: Complaints (Product)   61
Audit Approaches: Data Quality and Integrity   62
Audit Approaches: Laboratories (Microbiological,
    Analytical, and Development)   63
Audit Approaches: Vendor (Supplier)
    Requirements   64
Audit Approaches: Facilities and Equipment   67
Audit Approaches: Computer Systems   69

Audit Approaches: Commercialized Product   70
Audit Approaches: Investigational Product   71
Audit Approaches: General GMP Operations   72
Conclusion   76

7. Auditor Qualifications and Skills   **77**
Goals   77
What Is Required of a GMP Auditor?   78
Three Areas of Competence for Auditors   80
Ethics and Integrity   84
Becoming a Qualified Auditor   85
What to Expect as an Auditor   85
Conclusion   87

8. Phase I: Preparation   **89**
Goals   89
Who Is Involved   90
When It Is Done   90
Key Tasks   90
The Audit Plan   91
Milestones for Phase I: Preparation   107

9. Phase II: Conduct the Audit   **109**
Goals   109
Who Is Involved   110
When It Is Done   110
The Evaluation Process   110
The Desk Audit   111
The Site Audit   112
Site Audit Agenda   112
Milestones for Phase II: Conduct the Audit   124

10. Interviewing Auditees   **125**
Goals   125
What Is an Interview?   126
Interview Success   127

The First Critical Seconds of the Interview   127
Spoken Communication   128
Other Visual Cues   129
Conclusion   132

11. Phase III: Analyzing Audit Data   **133**
Goals   133
Who Is Involved   134
When It Is Done   134
Analysis While Collecting Data   135
Isolated vs. Systemic Deficiencies   137
Other Analytical Tools   140
Self-Analysis by Team Members   141
Milestones for Phase III: Analysis   141

12. Phase IV: The Audit Report   **143**
Goals   143
Who Is Involved   143
When It Is Done   144
Standard Formats   145
The Standard Audit Report   158
Working Papers   160
As You Write the Report . . .   160
Proprietary and Confidential Nature of the
    Information   161
Milestones for Phase IV: Report   162

13. Phase V: Follow-up and Closure   **163**
Goals   163
Who Is Involved   164
When It Is Done   164
What to Expect From the Auditee   165
Evaluating Auditee Responses   166
When Differences Arise . . .   166
Tracking Corrective Actions   168
Verifying Corrective Actions   168

Closure   170
Archiving Working Papers   170
Milestones for Phase V: Follow-up and Closure   170

14. Auditing Vendors and Contractors   **173**
Goals   173
Special Preparation for Vendor Audits   175
Conclusion   181

15. Being Audited   **183**
Goals   183
Phase I: Preparation   184
Phase II: The Audit   191
Phase III: Analysis   194
Phase IV: Report   194
Phase V: Following up   196
Conclusion   196

References and Readings   **197**
Appendix 1—American Society for Quality
    Control Code of Ethics   **199**
Appendix 2—In Their Own Words: GMP and
    Quality Auditors Talk About Auditing   **203**
Appendix 3—Resources   **215**
James L. Vesper   **219**
Index   **221**

──────────────── DEDICATION

To Noah Adam Vesper.

It's never too early to start.

PREFACE

The roots of this book extend back to the experiences and on-the-job training I had in the corporate Quality Assurance Division at Eli Lilly and Company. There, I had the opportunity of learning good manufacturing practice and quality auditing techniques from experienced colleagues. In the U.S., and overseas, I was able to apply these skills as we audited manufacturing and support sites.

The more recent catalyst for this book was being asked to create a workshop to teach auditing skills to pharmaceutical industry personnel. In designing the workshop, I talked with experienced auditors as well as auditees. I tried to understand the thought process used by auditors—something expert auditors do not consciously ponder; they simply audit.

Nobody is born an auditor—it requires basic knowledge and several different skill sets. To be a successful auditor, you need to:

- Have technical and regulatory knowledge and be able to apply it in various situations;
- Organize your thoughts and actions, following a defined process;

- Communicate effectively orally and in writing; and
- Work well with people.

This book is written for someone who is new to GMP auditing; it covers topics related to these four competencies, which will help make you successful. You will find the rationale for why auditing is an important quality tool, along with the audit cycle broken into five distinct phases.

The basic framework for auditing (the phases of preparation, conducting the audit, reporting, following-up) can be found in almost any book on the ISO 9000 quality system; in this book an additional phase, on analysis, is included.

Two other pieces that have been added are the *Types of Audit* and *Audit Approaches*. These are meant to give more structure to the questions, "What is the purpose of this audit?" and "What should I be looking for?" The categories and items presented in the *Type of Audits* and *Audit Approaches* are not meant to be definitive. They are guides that should be expanded upon as needed and kept current.

In Appendix 2, in the back of the book, are responses to a questionnaire from experienced auditors. Their personal insights are valuable for new auditors and will trigger memories for those who have been auditing for many years.

While this book is aimed at personnel new to GMP auditing, many of the concepts can be used in other areas. For example, currently, as I create GMP training programs and courses for clients, I do not formally audit firms. Instead, other training professionals and I conduct a *needs analysis* to define desired performance; examine actual performance; and identify the knowledge, skills, and performance tools/support that people need to do their jobs. This process has many elements in common with auditing; a background in GMP auditing has made this immensely easier for me.

To be successful at auditing you need more than this book—*you need to audit.* Working with other experienced auditors will help you apply the information from this book and allow you to develop the skills and instincts you need. Do not be alarmed if the seasoned auditor you work with doesn't have a detailed audit plan written up or a loose-leaf binder filled with checklists. When you have done dozens of audits every year for a decade, you may not need them either. Early in your career as an auditor, however, plans and checklists are tools you need to use and continually refine. You will find that the tools will not only help you audit, but will increase your professionalism and credibility with the audit sponsor and the auditee.

GMP auditing is a tremendous way to learn more about the drug and medical device industry. It also is important from the consumer/patient side. Every time you audit, you help assure that the products our families might use are safe, properly identified, pure, and of the highest possible quality. Please do a good job. We're depending on you.

Jim Vesper
May 1997

Acknowledgments

A number of people, unbeknownst to them, have contributed to this book. Naming people is always dangerous in case one forgets someone. In spite of that risk, thanks go to the GMP auditors I first worked with and from whom I learned so much: Ernie Barr, Tom Hebenstreit, Cline Mahoney, Jim Landes, and John White.

More recently, people who have been important during the writing of this book include colleagues and friends: Patricia Gatzios, Viveca Gavino, Janice Harrison, and Laura Pence. Other friends who have been encouraging and patient are Leslie Apetz, Robert Fink, Kimberly Fox, Sara Krusenstjerna, Harry Lang and Bonnie Meath-Lang, Jim Wanzenried, and Rob Williams. And, top honors go to Michael Thomas.

Initial book design concept and layout: Christopher M. Clark
Cover design: Kevin Southworth, Level 3, NTID Applied Art and Computer Graphics Sudent
Illustrations: Claire Marziotti

—————————CHAPTER 1

What Is a GMP or Quality Audit?

Goals

- Describe what an audit is.
- Provide the business reasons/benefits for conducting an audit.
- Provide the quality reasons/benefits for conducting an audit.
- Provide the compliance reasons/benefits for conducting an audit.
- Distinguish between internal and external audits; first, second, and third party audits.
- Describe what an audit is not.

Audit is one of those words we use that has a variety of meanings depending on its context. For students, *audit* means sitting in and observing a college course without receiving any credit. For Scientologists, audit is the process of freeing oneself of past painful experiences using an E-meter. For taxpayers, the threat of an audit causes an adrenaline surge and means finding receipts and trying to remember why certain decisions were made on their tax returns several years ago.

For people involved in the pharmaceutical or medical device industries who use the Good Manufacturing Practice (GMP) regulations, *audit* has other meanings, in some cases specifically defined by regulatory agencies such as the U.S. Food and Drug Administration (FDA) as shown in Table 1.1.

While there are financial, safety, environmental, and other forms of audits used in our industry, in this book we are focusing on one particular type: those used by pharmaceutical and medical device companies (and suppliers who provide components used in drugs and devices) to verify and evaluate adoption of the principles of good manufacturing practice and compliance to applicable GMP regulations. Different organizations give this different names: compliance auditing, quality auditing, GMP auditing, etc.

In this book, we will use one term, *GMP audit*, to encompass all of these and define it as

a documented, systematic tool, used by management and done periodically by independent, qualified people to verify and evaluate an organization's use of the principles of good manufacturing practice and compliance to applicable GMP regulations.

There is a great deal of meaning packed into that definition; it is worth taking it apart and examining it more carefully.

a documented, systematic tool . . .

Audits are one of several methods and approaches used by an organization to better understand itself or others. The audit is not an end but a way of reaching a particular goal. The activity and its results must be documented so

Table 1.1. Comparison of Definitions of *Audit*

Organization	Definition (source)
Canadian Health Protection Branch (HPB)*	A self-inspection programme appropriate to the type of operations of the company in respect to drugs, ensures compliance with Division 2, Part C of the [Canadian] Food and Drug Regulations. (4th Edition GMPs, Canadian HPB, 1996, Section C.02.012. (b) . . . The purpose of self-inspection is to evaluate the manufacturer's or importer's compliance with GMP in all aspects of production and quality control. The self-inspection programme is designed to detect any shortcomings in the implementation of GMP and to recommend the necessary corrective actions. (4th Edition GMPs, Canadian HPB, 1996, Section C.02.012. (b), Rationale)
European Community*	Self inspection: Should be conducted in order to monitor the implementation and compliance with Good Manufacturing Practice principles and to propose necessary corrective actions. (European Community [EC] Guidelines, 1994, Section 9—Self Inspection.)
International Organization for Standardization (ISO)/American Society for Quality Control (ASQC)	Quality Audit: A systematic and independent examination to determine whether quality activities and related results comply with planned arrangements and whether these arrangements are implemented effectively and are suitable to achieve objectives. Notes: 1. The quality audit typically applies to, but is not limited to, a quality system or elements thereof, to processes, to products, or to services . . .

Continued on next page.

Continued from previous page.

Continued from previous page.

	2. Quality audits are carried out by staff not having direct responsibility in the areas being audited but, preferably, working in cooperation with the relevant personnel. 3. One purpose of the quality audit is to evaluate the need for improvement or corrective action. An audit should not be confused with surveillance or inspection activities performed for the sole purpose of process control or product acceptance. 4. Quality audits can be conducted for internal or external purposes. (ANSI/ISO/ASQC Q100011-11994, 1994. Guidelines for Auditing Quality Systems—Auditing.)
U.S. FDA	Quality Audit: a systematic, independent examination of a manufacturer's quality system that is performed at defined internals and at sufficient frequency to determine whether both quality system activities and the results of such activities comply with quality system procedures that these procedures are implemented effectively, and that these procedures are suitable to achieve quality system objectives. (Quality System Regulations, 1996, 21 CFR 820.2 (t))**

* While the agencies define other terms in the definition or glossary portions of their requirements, neither specifically defines GMP audits or their term *self-inspections*.
**Also referred to as the "Medical Device GMPs."

it can be shared with those who can take action based on the audit's findings. The audit is also done in a planned, organized way so potentially critical issues are not passed over or swept aside.

. . .used by management . . .

The audit helps management see how well (or poorly) their philosophy, policies, and mandates are understood and put into practice.

. . .done periodically . . .

Audits give management a view from a particular vantage point. Depending on how it is done, it could be "snapshot" in time, or a retrospective view looking at documentation and records. While there may be some reasons for doing an audit only once, having different views over several years allows management to see the auditee in another dimension: over time. This way, improvement can be monitored.

. . .by independent, qualified people . . .

For the process and results to be credible, those doing the audit need to be free of organizational bias. They must not be out to "get" anybody or have a preconceived notion of the audit results. Auditors are there to gather data in a scientifically defendable way and then to make judgments based on the data. Not only do GMP auditors need a strong background in auditing itself, they also need a practical point of view to discover quickly if there are any skeletons buried, and if there are, what the quality, business, and regulatory implications may be. If you don't have confidence in the people doing the audit, there is no way you will have confidence in the results.

. . .to verify and evaluate . . .

The audit is not an inspection: it determines if quality system elements exist, if they meet the regulatory/

organization requirements and expectations, and how well they are working in actual practice.

. . .an organization's use . . .

Auditors don't just look at what a firm has down on paper in its policies and procedures. Auditors determine if the policies and procedures have been implemented. A firm may have procedure binders with its name embossed in gold leaf, but if they are not current or being used or if the performance differs from what is described by the procedure, you have a problem.

. . .principles of good manufacturing practice and compliance to applicable GMP regulations.

These are the standards to which the firm is compared. Since the GMPs are predominantly performance standards, it is up to each firm to accomplish the goals as it sees fit. An organization with a deep understanding of GMP principles will not only meet the regulatory requirements but exceed them. Such an organization will reduce the many gray areas so that the organization will not just comply with the regulations but thrive in a highly regulated environment. One of the key reasons that qualified auditors are needed is they must interpret the requirements and how they should apply in a particular situation.

Why Conduct GMP Audits?

With an understanding of what a GMP audit is all about, the next question you might ask is, why do it? What value is there to all of this work?

One of the most practical reasons for auditing came from an executive of a large U.S. pharmaceutical company.

He said that audits help him sleep better at night. Even though there might be problems within parts of the organization, at least he is aware of them and plans have been made to address the deficiencies. There are a number of other benefits in doing a GMP audit. Here are a few of them; if you spend a few minutes thinking about it, you will easily come up with many more.

- **Benchmark of current performance:** An audit will show how activities and existing systems are being used at the time of the audit. This can be a baseline against which future improvements are compared.
- **Better deployment/redeployment of resources:** The audit results may show that people need to be reassigned and expenditures need to be made in particular areas. In some situations, this could result in a savings to the organization.
- **Standardization:** Applying the same expectations at various sites or departments can help achieve consistent performance throughout an organization.
- **Avoiding problems:** Auditing potential vendors helps to minimize the chances of "if we only had known. . . ." Having a structured GMP auditing program and carefully identifying what to look for will help to avoid or minimize problems that could occur later.
- **Means for improvement:** Identifying problems is the first step in solving them; audits can do this. In addition, the audit may point out the root and contributing causes of the problems.
- **Communication:** Different types of communication occur as part of the audit. Standards/expectations are communicated when checklists are shared with the auditee; management concern is shown by the fact that there is an audit. Ideas and

solutions to common problems are shared between the auditors and auditees.

- **Training:** The auditee's primary contact and all those working with the auditor have an opportunity to learn the auditor's view of how GMP principles and regulations apply. A qualified auditor should be able to explain the rationale behind requirements and how the auditee can use them to add value to its operation. From the auditor's perspective, each audit is a chance to learn about an organization and its approach in meeting GMP expectations.
- **Compliance:** If the audit is done properly, compliance will be only one of the many benefits. Management that makes compliance the primary focus of the audit shows a lack of understanding of an audit's power.

Terms Related to GMP Audits

In its most basic form, an audit involves three groups:

1. **Sponsor** (or client)—the organization that wants the audit conducted
2. **Auditors**—those conducting the audit
3. **Auditee**—the organization being audited

How these three groups relate to each other in an audit is described by another set of terms. When the sponsor, auditor, and auditee are *all part of the same organization*, it is considered an **internal audit** or **first-party audit** (Figure 1.1).

When the sponsor and auditor are part of one organization and the auditee is part of another organization, it is considered an **external audit** (Figure 1.2). There are several variations on this as well.

Figure 1.1. Example of a first-party audit.

Figure 1.2. Example of an external audit.

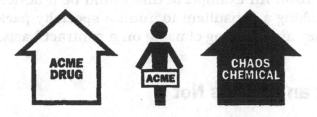

Figure 1.3. Example of a second-party audit conducted by a regulatory agency.

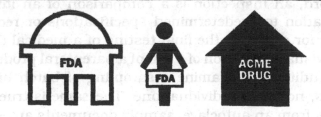

Figure 1.3 shows a **second-party audit**—the sponsor and the auditor are from the same organization. Regulatory audits are of this type: the agency (e.g., FDA, HPB) would be the sponsor and provide the auditor (i.e., inspector), who examines the auditee.

Figure 1.4 illustrates a **third-party audit**. Here, there is only a limited connection between the sponsor and the auditor who will be examining the auditee on behalf of

---

Figure 1.4. Example of a third-party audit.

ACME DRUG · INDEPENDENT CONSULTANT · CHAOS CHEMICAL

---

the sponsor. An example of this would be a device company asking a consultant to audit a specialty packaging firm they are thinking of using on a contract basis.

What an Audit Is Not

Some people have misconceptions about auditing, in part due to how the term has been misapplied. For example, an audit is *not* truly an inspection. In the strict sense of the word, an inspection is a comparison of an item or information to predetermined specifications or requirements, for example, the final testing of a medical device or the visual inspection of a vial of a parenteral product. A GMP auditor will examine a sampling of batch history records, not each individual one. The same is true with records from an autoclave: sample documents are examined and any observations made are based on those samples. Inspection, on the other hand, usually involves someone looking at each and every item and making a judgment about it. Despite their titles, *Pre-Approval Inspections* and *Post-Approval Inspections* are really audits.

Audits are *not* the judgments made to release a batch of product. The process of determining whether or not the product should be released could, however, be the focus of a GMP audit (to be sure all documents were properly reviewed, any deviations were resolved, etc.).

Finally, a GMP audit is *not* the informal inspection that Quality Control (QC) might conduct or that a department does on its own. These are very useful monitoring tools, but they seldom have the rigor, formal organizational support, and impartial observers found in a GMP audit.

Conclusion

A GMP audit is a useful, powerful tool that, when conducted by qualified personnel, can provide a wide range of benefits to the audit sponsor as well as the auditee.

———————————————— CHAPTER 2

The Big Picture

Goals

- Identify the audit as part of a system.
- Describe the five phases of conducting an audit.
- Describe what happens in each phase.
- Discuss the connection to past and future audits.
- Discuss the outcome of an audit.

The Audit as a Quality System Component

GMP audits are a component of a drug or device firm's quality system. That means it is part of an organization's structure, responsibilities, components, procedures, and resources that are integrated to meet the quality and business goals of the firm and its customers.

While some quality system components are of a baseline variety (i.e., they are used for basic, normal operations) and others are required for change and problem management, GMP audits are used to monitor and improve products and operations (Table 2.1). GMP audits can look at all aspects of a firm's compliance strategy and quality system. Conducting these audits and using their findings is essential to continually improving products, processes, information, and services. .

Five Phases of the Audit Process

Auditing includes five phases (Figure 2.1). When you complete all five phases, you have gone through one audit cycle. In later chapters, each phase and the tasks that make up each phase will be discussed in detail.

Phase I: Preparation

Preparation of an audit includes developing the audit plan, identifying what will be examined, arranging the details of the audit, and communicating with management and auditee.

The tasks comprising Phase I can easily be the most time consuming of the audit cycle, particularly if there are no auditing tools, such as checklists, standardized audit plan sheets, etc. available to the auditors. In the initial stages of preparation, you will be choosing the type of audit to be conducted, as well as the specific audit approach(es) that you will use. These decisions are based primarily on the goal of audit. Knowing this information will permit you to develop checklists and action plans from suggestions found in Chapter 5.

Typically, the preparation phase is approximately 2 to 3 days for every day spent auditing.

Table 2.1. Examples of Quality System Components

Category	Component
Baseline	Batch Release
	Training
	Validation
	Preventive Maintenance
	Calibration
Problem and change management	Change Control
	Deviation Investigation
	Notifications to Management
	Product Recall
Monitoring and improvement	GMP Auditing
	Stability
	Periodic Production
	Quality Reviews
	Sample Retention

Figure 2.1. Five phases of the audit process.

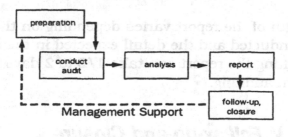

Phase II: Conduct the Audit

Conducting the audit means implementing the audit plan by reviewing documentation and visiting the site, collecting data, making observations, and presenting a preliminary, oral report to the auditees.

The time required for this phase can range from one-half day for auditing a small vendor to several weeks when conducting a comprehensive audit of a facility. Normally, because of time and resource limitations, audits rarely last longer than two weeks.

Phase III: Analysis

Analysis of the audit means organizing and obtaining meaning from the data and the information collected.

The time spent analyzing the audit results takes approximately 1/4 to 1 day for every day spent auditing.

Phase IV: Report

The report phase of an audit means preparing and distributing a completed report to management.

The length of the report varies depending on the type of audit conducted and the detail expected in the audit report. Writing the report can take 1/2 to 2 days for every day spent auditing.

Phase V: Follow-up and Closure

Follow-up and closure of an audit means confirming that the auditee's response to the report did, in fact address the audit's issues, and that commitments were put into action.

The amount of time involved in following up an audit depends on how many observations were noted and the confidence the auditors (and their management) have in the commitments being accomplished. If a follow-up is made without another site visit, it may take 1/4 to 1 day per auditing day, accumulated over several months, to bring the audit to closure.

Management Support

One other aspect of the auditing program is management's use of the auditing data in monitoring the organization on a macro level. Some firms consider results of GMP audits on par with financial audits, reviewing formal GMP audit reports at Board of Director meetings. Consistency in GMP auditing approaches and standards can make these reports and comparisons more useful.

Monitoring Improvement

One way that audit results are used is to measure change over time. The notion is linked to the term *audit cycle*, which implies an audit is not done just once, but repeatedly.

As will be discussed in Chapter 12, audit reports provide management and other auditors a picture of the auditee's operations and quality system at one point in time. If there is some comparability of the audit documents, putting these pictures side-by-side can show trends.

Outcomes of the Audit

Apart from the chance to get away from the office and an occasional road trip, GMP audits are done with a larger purpose. Audits are a tool, a method to get from one place to another. You don't audit just to audit; you audit to help make a decision, to monitor, to assure, to improve.

Having a clear, commonly understood goal is critical to the success of the audit. Quality Assurance (QA) auditors have told stories about being sent to conduct a routine survey of a supplier when their management *really* wanted to get better quality components from the supplier. If all parties had clearly stated/understood that the goal of the audit was to "evaluate the vendor's quality systems and identify ways of achieving consistent conformance to specifications," the audit would be more useful to all concerned.

Probably the most important point in the audit occurs when the auditee receives the final report. Since auditing can be a tool for improvement, the oral and written reports need to set a course for the improvement. Auditors need to be specific as to what the issue is and whether the finding is an isolated occurrence or a systematic failure; the corrective actions in each case can be very different.

Conclusion

GMP audits are one component of a firm's quality system. Audits are more than a casual, quick visit to an area; they are carefully planned and executed by skilled personnel. The information that results from an audit not only helps management make critical decisions, but can assist the auditee as they monitor and improve their operations.

CHAPTER 3

— CHAPTER 3

Regulatory Requirements and Expectations for GMP Audits

Goals

- Identify specific instances when GMP regulations require GMP-type audits.
- Identify what needs to be included in agency-required audits.
- Distinguish between GMP regulatory requirements and expectations.

In Chapters 1 and 2, we examined what a GMP audit is and what it consists of from a quality system point of view. We also saw the different phases and activities that make up an audit cycle. In this chapter, we will see what different regulatory bodies require or expect in terms of GMP auditing.

Compliance should not be the force driving GMP audits. If it is, the firm's management has a limited view of an audit's benefits. However, if an audit is not done with an understanding of the regulatory expectations, it will not meet one of its purposes.

Requirements vs. Expectations

The cGMP regulations issued by the U.S. FDA (i.e., the Current Good Manufacturing Practice Regulations) include, literally, more than what is found in the Code of Federal Regulations (i.e., 21 CFR). What is written in Parts 210 and 211 for the drug industry and Part 820 for the medical device industry are the *minimum requirements*. Industry practice, FDA concerns (as published for its investigators and industry as "Guidance" documents), new technology, etc. can raise the expectations that inspectors have as they examine an operation.

U.S. cGMPs for Finished Pharmaceuticals (21 CFR Parts 210 and 211)

There is no explicit requirement for internal GMP or quality audits by a drug manufacturer of itself. There is, however, an expectation that a firm would have such a system in place, based on industry practice.

For vendors (or contract acceptors) of raw materials or packaging components there is no explicit requirement of a GMP audit. Again, however, industry practice is to conduct an initial audit of a proposed vendor and then periodic reaudits of the vendor.

One way that the FDA puts teeth into the expectation of having approved, audited vendors is in the submission process. Firms are required to list the sources of critical materials, such as active pharmaceutical ingredients (i.e., bulk drug substances); these sources are inspected by FDA personnel as part of the New Drug Application (NDA) or Abbreviated New Drug Application (ANDA) Pre-Approval Inspection Program.

Could your firm get cited by the FDA for not having a vendor auditing program even though it is not specifically required by the drug cGMPs? Probably yes, particularly if there was an indication that deviations or quality problems might be related to the materials supplied by a vendor. The investigator could do this by relying on current industry practice or by indicating that your firm has no real assurance about the identity, quality, and source of the products it is receiving. An inspector could start off asking the question, "How do you know. . . ?" Without having first-hand knowledge about the vendor's practices, it would be hard to effectively answer the question. Even more important, however, are the business reasons for having a systematic way of evaluating and approving suppliers. The quality of the products you create depends on the quality of materials used to manufacture them.

U.S. cGMPs for Medical Devices: Quality System Regulation (21 CFR Part 820) [Published October 7, 1996]

The new FDA medical device regulations, written to better harmonize with ISO-9000, require that firms conduct quality (GMP) audits.

§ 820.22 Quality audit.
Each manufacturer shall establish procedures for quality audits and conduct such audits to assure that the quality system is in compliance with the established quality system requirements and to determine the effectiveness of the quality system. Quality audits shall be conducted by individuals who do not have direct responsibility for the matters being audited. Corrective action(s), including a reaudit of deficient matters, shall

be taken when necessary. A report of the results of each quality audit, and reaudit(s) where taken, shall be made and such reports shall be reviewed by management having responsibility for the matters audited. The dates and results of quality audits and reaudits shall be documented.

Some additional detail about the FDA's "original intent" is provided in the "Preamble" to the regulations, also published on Oct 7, 1996:

> 53. A few comments stated that § 820.20(c) should be deleted because it duplicates the quality audit required by § 820.22. FDA disagrees that § 820.20(c) duplicates the requirements in § 820.22. The purpose of the management reviews required by § 820.20(c) is to determine if the manufacturer's quality policy and quality objectives are being met, and to ensure the continued suitability and effectiveness of the quality system. An evaluation of the findings of internal and supplier audits should be included in the § 820.20(c) evaluation. The management review may include a review of the following: (1) The organizational structure, including the adequacy of staffing and resources; (2) the quality of the finished device in relation to the quality objectives; (3) combined information based on purchaser feedback, internal feedback (such as results of internal audits), process performance, product (including servicing) performance, among other things; and (4) internal audit results and corrective and preventive actions taken. Management reviews should include considerations for updating the quality system in relation to changes brought about by new technologies, quality concepts, market strategies, and other social or environmental conditions. Management should also review periodically the appropriateness of the review frequency, based on the findings of previous reviews. The quality system review process in § 820.20(c), and the reasons for the review, should be understood by the organization. The

requirements under § 820.22 Quality audit are for an internal audit and review of the quality system to verify compliance with the quality system regulation. The review and evaluations under § 820.22 are very focused. During the internal quality audit, the manufacturer should review all procedures to ensure adequacy and compliance with the regulation, and determine whether the procedures are being effectively implemented at all times. In contrast, as noted above, the management review under § 820.20(c) is a broader review of the organization as a whole to ensure that the quality policy is implemented and the quality objectives are met. The reviews of the quality policy and objectives (§ 820.20(c)) should be carried out by top management, and the review of supporting activities (§ 820.22) should be carried out by management with executive responsibility for quality and other appropriate members of management, utilizing competent personnel as decided on by the management.

55. A few comments suggested that FDA delete the requirement that persons conducting the audit be "appropriately trained" from the second sentence of proposed § 820.22(a), because it is subjective and not consistent with ISO 9001. FDA has deleted the requirement from § 820.22(a) because § 820.25 Personnel requires that such individuals be appropriately trained. Further, FDA has attempted to better harmonize with ISO 9001:1994, which does not explicitly state personnel qualifications in each provision. Similarly, in response to general comments suggesting better harmonization, FDA has added the requirement that the audit "determine the effectiveness of the quality system" as required by ISO 9001:1994. This requirement underscores that the quality audit must not only determine whether the manufacturer's requirements are being carried out, but whether the requirements themselves are adequate.

56. Some comments stated that requiring "individuals who do not have direct responsibility for the matters

being audited" to conduct the audits is impractical and burdensome, particularly for small manufacturers. FDA disagrees with the comments. Both small and large manufacturers have been subject to the identical requirement since 1978 and FDA knows of no hardship, on small or large manufacturers, as a result. Small manufacturers must generally establish independence, even if it means hiring outside auditors, because the failure to have an independent auditor could result in an ineffective audit. Manufacturers must realize that conducting effective quality audits is crucial. Without the feedback provided by the quality audit and other information sources, such as complaints and service records, manufacturers operate in an open loop system with no assurance that the process used to design and produce devices is operating in a state of control. ISO 9001:1994 has the same requirement for independence from the activity being audited.

57. Several comments claimed that the last sentence in proposed § 820.22(a), which required that follow-up corrective action be documented in the audit report, made no sense. The comments said that corrective action would be the subject of a follow-up report. It was the agency's intent that the provision require that where corrective action was necessary, it would be taken and documented in a reaudit report. The provision has been rewritten to make that clear. New § 824.22 also clarifies that a reaudit is not always required, but where it is indicated, it must be conducted. The report should verify that corrective action was implemented and effective. Because FDA does not review these reports, the date on which the audit and reaudit were performed must be documented and will be subject to FDA review. The revised reaudit provision is consistent with ISO 9001:1994.

58. Many comments were received on proposed § 820.22(b) regarding the reports exempt from FDA

review. Most of the comments objected to FDA reviewing evaluations of suppliers. FDA has decided not to review such evaluations at this time and will revisit this decision after the agency gains sufficient experience with the new requirement to determine its effectiveness. A thorough response to the comments is found with the agency's response to other comments received on § 820.50 Purchasing controls. FDA has moved the section regarding which reports the agency will refrain from reviewing from § 820.22(b) to new § 820.180(c), "Exemptions," under the related records requirements. FDA believes this organization is easier to follow.

There is no explicit requirement in the FDA's medical device Quality System Regulations for conducting audits of vendors; however, an audit would be one way of evaluating the supplier.

§ 820.50 Purchasing controls.
Each manufacturer shall establish and maintain procedures to ensure that all purchased or otherwise received product and services conform to specified requirements. (a) Evaluation of suppliers, contractors, and consultants. Each manufacturer shall establish and maintain the requirements, including quality requirements, that must be met by suppliers, contractors, and consultants. Each manufacturer shall: (1) Evaluate and select potential suppliers, contractors, and consultants on the basis of their ability to meet specified requirements, including quality requirements. The evaluation shall be documented.

Again, the Preamble gives some idea of what the agency was thinking as they wrote this section on suppliers and purchasing controls:

103. One comment stated that many suppliers of components to the medical device industry have their quality

systems certified to an ISO 9000 standard by an independent third party auditor, and that such registration of component manufacturers should be considered in vendor assessment plans. FDA agrees in part with the comment in that certification may play a role in evaluating suppliers, but cautions manufacturers against relying solely on certification by third parties as evidence that suppliers have the capability to provide quality products or services. FDA has found during inspections that some manufacturers who have been certified to the ISO standards have not had acceptable problem identification and corrective action programs. Therefore, the initial assessment or evaluation, depending on the type and potential effect on device quality of the product or service, should be a combination of assessment methods, to possibly include third party or product certification. However, third party certification should not be relied on exclusively in initially evaluating a supplier. If a device manufacturer has established confidence in the supplier's ability to provide acceptable products or services, certification with test data may be acceptable. Section 820.80 is specific to a device manufacturer's acceptance program. While finished device manufacturers are required to assess the capability of suppliers, contractors, and consultants to provide quality products and services, inspections and tests and other verification tools are also an important part of ensuring that components and finished devices conform to approved specifications. The extent of incoming acceptance activities can be based, in part, on the degree to which the supplier has demonstrated a capability to provide quality products or services. An appropriate product and services quality assurance program includes a combination of assessment techniques, including inspection and test.

107. Several comments stated that it was not clear how a manufacturer could evaluate an off-the-shelf component that is purchased from a distributor rather than

directly from its manufacturer, and stated that it would not be helpful to audit the distributor. FDA agrees that auditing a distributor would not meet the intent of § 820.50. Manufacturers should remember that the purpose of assessing the capability of suppliers is to provide quality products and to provide a greater degree of assurance, beyond that provided by receiving inspection and test, that the products received meet the finished device manufacturer's requirements. The agency recognizes that finished device manufacturers may not always be able to audit the supplier of a product. In such cases, the manufacturer must apply other effective means to assure that products are acceptable for use.

Additional information about how FDA inspectors are told to examine an audit program, policies, and procedures can be found in Chapter 4.

Canadian Requirements

The Canadian Health Protection Branch (HPB) is the agency responsible for issuing the GMPs in Canada. Several sections in their GMPs that discuss GMP audits or self inspections are provided below:

C.02.012.
Every manufacturer and importer of a drug shall maintain . . .(b) a program of self inspection.

Interpretation
2   A self-inspection programme appropriate to the type of operations of the company, in respect to drugs, ensures compliance with Division 2, Part C of the Food and Drug Regulations.
2.1 A comprehensive written procedure is available and describes the functions of the self-inspection programme.

2.2 The programme of a manufacturer engaged in processing a drug from raw material through to the drug's dosage form addresses itself to all aspects of the operation. When manufacturers are engaged only in packaging and/or distributing drugs produced by another manufacturer, the written programme covers only those aspects of the operations over which the manufacturer exercises control on his premises.

2.3 The self-inspection team includes staff and others who are suitably trained and qualified in GMP.

2.4 Periodic self-inspections are carried out.

2.5 Reports on the findings of the inspections and corrective actions are reviewed by appropriate senior company management. Corrective actions are implemented in a timely manner.

3   To ensure compliance of contractors and suppliers...

3.2.2 The contract or agreement permits the manufacturer or importer to audit the facilities of the contract producer.

Other sections of the Canadian GMPs discuss auditing as part of a supplier certification program: Raw Materials Testing. (Note: a similar requirement is found under Packaging Materials Testing, Section C.02.017.)

C.02.010

Interpretation

2. A raw material supplier certification programme, if employed, is documented in a standard operating procedure. There are several acceptable approaches to supplier certification. Such a programme may include, amongst others, the following auditing criteria:

. . .2.2 an on-site audit of the supplier's facilities and controls by a qualified person using acceptable criteria . . .

European Community (EC) GMP Guidelines

Several references to internal audits found in the EC GMP Guidelines include the following:

1 Quality Management – Quality Assurance
1.2 ix. There is a procedure for Self-Inspection and/or quality audit which regularly appraises the effectiveness and applicability of the Quality Assurance system.

7 Contract Manufacture and Analysis
7.14 The contract should permit the Contract Giver to visit the facilities of the Contract Acceptor.

9 Self-Inspection
Self-inspections should be conducted in order to monitor implementation and compliance with Good Manufacturing Practice principles and to propose necessary corrective measures.
9.1 Personnel matters, premises, equipment, documentation, production quality control, distribution of medicinal products, arrangements for dealing with complaints and recalls, and self-inspection, should be examined at intervals following a pre-arranged programme in order to verify their conformity with the principles of Quality Assurance.
9.2 Self-inspections should be conducted in an independent and detailed way by designated competent person(s) from the company. Independent audits by external experts may also be useful.
9.3 All self-inspections should be recorded. Reports should contain all observations made during the inspections and, where applicable, proposals for corrective measures. Statements on the actions subsequently taken should be recorded.

Conclusion

GMP auditing strictly for compliance reasons should not be the primary reason for a firm's GMP auditing program; however, GMP requirements and expectations should be known and met for the auditing activities to have maximum benefit to the organization.

CHAPTER 4

Your Auditing Policy and SOP

Goals

- Distinguish between a policy and procedure (SOP).
- Identify what should be included in a GMP auditing policy.
- Identify what should be included in a GMP auditing SOP.
- Discuss how regulatory requirements affect what should be in a GMP auditing policy and procedure.

In Chapter 2, we saw that GMP auditing was one of several components in a GMP/Quality System. Many firms define this system in policies found in their quality manual. Firms certified under ISO-9000 or medical device manufacturers regulated by the U.S. FDA are required to have quality manuals. In addition to the policies covering GMP auditing, firms usually have standard operating procedures (SOPs) that provide more detail on the auditing process. Some firms also have internal auditing handbooks that provide additional background information, forms, and data collection tools to auditors.

Policies and Procedures

Policies are different from procedures; it is common for firms to have both. Policies are documents that describe the who, the what, and the why. They are a high-level document in terms of the detail provided, as there is very little information about how you actually accomplish a task. For example, a policy on GMP auditing would cover the following:

- Who is responsible for GMP auditing within the organization,
- What a GMP audit includes,
- What the different types of GMP audits are,
- What happens with the audit report, and
- Why GMP audits are done.

Procedures are mid-level documents that emphasize the what and the how; they contain more detail than policies. A firm might have several SOPs related to auditing, one covering supplier audits, another on internal audits, etc., with each SOP providing instructions to the step or substep level. The SOP for auditing a supplier might cover the following:

- When to audit the supplier;
- Who specifically conducts the audit;
- The tasks involved in auditing (e.g., the phases);
- What documents the auditors produce;
- What the allowed timeframes are for certain activities (e.g., issuing the audit report);
- Who receives a copy of the audit report;

- How the audit report is reviewed, approved, and distributed; and
- How to archive audit documents and working papers.

Your auditing policy and SOP(s) need to address specifically any regulatory requirements, such as how often an audit is to be conducted, approval of the audit reports, who is qualified to be an auditor, etc. These requirements are determined by a careful reading of the GMP regulations that apply to your organization. (See Chapter 3 for examples of some specific requirements.)

One other type of document found in documentation systems, especially those following an ISO-9000 approach, is *work instructions*. Work instructions are highly detailed, prescriptive documents that emphasize the steps and substeps that are taken when performing a specific task. For a particular GMP audit, the work instructions would include the audit plan and the audit action plan. These are considered *working papers* and are discussed in more detail in Chapter 8.

The policies and procedures that have been approved and put into effect determine the way your firm operates and how GMP audits are conducted. Unapproved changes, as in other GMP contexts, would be considered deviations.

Regulatory Inspection of GMP Auditing Policies and SOPs

The U.S. FDA considers internal auditing reports to be confidential information and, according to FDA policy CPG 7151.02, will not typically ask to review such reports. They can, however, examine the auditing program that is required by the regulations (see Chapter 3), as defined in your own policies and procedures.

The FDA's *Investigations Operations Manual* (FDA 1996B) tells FDA inspectors what they should look for when they examine a medical device firm's quality audit program:

. . . If significant quality assurance problems have existed both before and after the firm's last self-audit, then you should critically review the written audit procedures. The audit procedures should cover each quality assurance area, and should be specific enough to enable the person conducting the audit to perform an adequate audit. If it is possible to interview an auditor, ask how the audits are performed; what documents are examined; how long audits take; etc.

Audits should be conducted by individuals not having direct responsibility for matters being audited. In one-person and very small firms, where hiring an outside auditor to meet this requirement would be impractical or overly burdensome, self-audit may be acceptable and the auditor may not be independent.

. . . If there are significant FDA-483 observations, and independent audits are being performed, but deficiencies are apparently not being identified by the auditor, then an FDA-483 should contain an observation indicating a lack of adequate audits.

If possible, attempt to determine whether corrective action by upper management is being taken. Auditors may be asked if they observed any of the ongoing GMP deficiencies during their prior audits (ongoing GMP deficiencies may also be identified by reviewing prior FDA-483's). If the answer is yes, check the written audit schedule to determine if a follow-up audit is scheduled for the deficient areas. Also, check the written audit procedure for instructions for review of audits by upper management, and re-audit of deficient areas. A failure to implement follow-up corrective actions may be listed as a GMP deficiency on the FDA-483.

There are several points, then, that should be included in your SOP:

- Definition of areas to be audited,
- How the audits are to be conducted,
- Qualifications/training required for an auditor,
- Organizational area responsible for auditing and providing the auditor,
- When audits are to be conducted,
- Who is responsible for following-up on audit observations, and
- When reaudits are required for closing out an audit.

Certification of an Audit

The U.S. FDA's Quality System Regulations for medical devices (i.e., the medical device cGMPs) have a provision allowing an FDA inspector to request that the firm's management certify in writing that an audit has been conducted and that any corrective actions have been taken. As with other certifications that the FDA asks management to make, this carries with it a danger: The FDA, backed by the U.S. Supreme Court,

> . . . will prosecute corporate officials even when they are not directly responsible for the violations cited.
>
> The personal liability company employees face is one reason for the strong resistance to FDA's increasingly common demand in GMP complaints that top executives "certify" that appropriate corrective actions have been taken. The false certification to the government, whether deliberate or by error, can trigger criminal prosecution even without a violation of the FD&C Act (GMP Letter 1996).

Regulatory Inspection of Vendor Audits

The U.S. medical device Quality System *does not* consider audits of vendors or contractors as privileged documents. That is,

> Audit reports written as part of the assessment of suppliers or contractors (§ 820.50(a)) are subject to review and copying by the FDA (FDA 1996A).

Figure 4.1 is an example of what a drug or medical device firm might have as an auditing policy.

Figure 4.1. Example of a policy on GMP auditing.

Omega Health Products	Title:	GMP Auditing
	ID No.	QP101.2
Corporate Quality Policy	Effective Date:	Oct 1/96
	Approval:	QCom, Aug 96

Goal	Periodically evaluate facilities to determine their success at meeting local regulatory standards (GMPs) and corporate quality requirements.
Scope	This Quality Policy applies to all OHP companies and subsidiaries worldwide.
Policy	1. OHP Corporate Quality Assurance (CQA) is responsible for establishing a program to systematically evaluate all GMP facilities on a regular basis. 2. Expectations, standards, and references will be developed, published, and distributed by CQA to sites at least six months prior to the scheduled audit date.

Continued on next page.

Continued from previous page.

3. Each year, in the fourth quarter, CQA will develop a Proposed Audit Schedule for the upcoming year. The corporate Quality Committee will discuss and approve this at their November or December meeting.

4. Audit dates must be agreed to by site management and CQA.

5. CQA will lead all audits and include at least one auditor from Technical Operations.

6. At the end of each GMP Audit, the audit team will provide site management with a preliminary oral report of their findings and its significance.

7. Draft reports will be reviewed and commented upon by CQA and the audited site's director.

8. Completed audit reports will be provided to the following within 20 business days following the audit.
- Site Tech Ops Director
- Site Quality Director
- Director of CQA
- VP of Quality
- Executive VP of Operations

9. When there are audit findings requiring corrective actions, these actions shall be documented and approved by the site directors within 20 business days. Time frames for corrective actions are permitted. Actions plans are to be addressed to the Director of CQA.

10. CQA will follow up on the corrective actions. When completed, CQA will publish an Audit Closure Memo.

11. The Quality Committee will monitor CQA auditing activities and site corrective action plans.

Conclusion

Policies and procedures that state the requirements and methods for GMP audits are important not only because they set the expectations. These documents, because they are reviewed and approved by various units of the organization, also communicate management's endorsement of audits as a tool to monitor and improve quality. Procedures that are well thought out and have broad organizational support make it easier for auditors to be efficient and effective.

—————————————CHAPTER 5

Types of Audits

Goals

- Describe a *type of audit*.
- Describe different types of internal and external audits.
- Discuss how the goal of the audit relates to the type of audit that is selected.
- Discuss when to use a particular type of audit.

Types of Audits

In the first chapter, we saw that GMP audits are categorized into the following:

- Internal
  —first party
- External
  —second party
  —third party

The category of audit depends on the relationship between the auditor and the auditee. These terms are useful in discussing organizational aspects of the audit.

When talking about the *purpose* of the audit, it is more useful to talk about the type of GMP audit that is going to be conducted. This concept is closely linked to the goal or desired outcome of the audit. The *goal* is why an audit is being conducted. What difference will the audit make to the audit's sponsor and the auditee? This basic question *must* be answered before the audit is conducted.

The type of audit that is selected can be performed internally by the firm's own auditors (i.e., first party), or externally, with auditors from a contract giver, a consulting firm, or regulatory agency.

Examples of Types of GMP Audits

Below is a listing of types of audits; they will be described in more detail later in this chapter.

- Informal
- Quality or Management
- Supplier or Contractor
- Focused
- Preparatory
- Benchmarking
- Due diligence
- Certification
- Regulatory

Making Use of Various Types of Audits

Once you know the organization you will be auditing and you understand the goal of the audit, you can find a type of audit that best matches your situation. Then, using the description and goal for the type of audit, you can select or create the audit approaches (Chapter 6) that will help accomplish the goal.

Informal Inspections

Description: A periodic examination of the general operations, quality systems, and GMP compliance of an area.

Goal: Discover and correct deficiencies.

Who conducts: Department or quality unit personnel.

Comments: Informal inspections are a tool used to maintain and improve general GMP compliance. Sometimes these are called *self-inspections* or *self-audits*, because department personnel take a look at their own activities using standard checklists or forms. Some organizations link this type of self-examination to their training efforts. For example, a training session on pest control would include an activity where teams evaluate their areas for vulnerabilities discussed in the training.

While informal inspections can be an important part of a continual improvement process, *an informal inspection is not a GMP audit* in the strict sense of the phrase, because auditors from the department are not really impartial. Also, documentation of the event typically does not include a formal report and closure.

Quality or Management

Description: An examination of the systems, practices, equipment, and facilities used in developing and producing products and related information.

Goal: Determine how well the firm is meeting its quality, compliance, and business objectives.

Who conducts: Can include internal auditors (e.g., Quality Assurance) or external auditors (e.g., customers or consultants).

Comments: These are usually done on the mandate of a corporate group, such as Quality Assurance, for example, a team from corporate Quality Assurance visiting one of their company's manufacturing sites.

Supplier or Contractor

Description: An examination of a proposed or current supplier of materials/components, or a firm contracted to manufacture, package, label, or test a product.

Goal: Determine if the supplier or contractor is suitable to be (or remain) qualified as a vendor. Also, to determine if a current supplier is meeting its contractual commitments.

Who conducts: Typically a team, including individuals from Quality Assurance, Quality Control, Technical Operations, or Purchasing of the customer firm (i.e., the contract giver). Sometimes this activity is subcontracted to a third-party auditor.

Comments: This is probably the type of audit that is done most frequently, with customers periodically visiting their suppliers and potential suppliers. This can be one of the more difficult auditing situations, as vendors may receive audits from many customers, each wanting the vendor to meet their own requirements and expectations. (For more information, see Chapter 14, Auditing Vendors and Contractors.)

Focused

Description: An in-depth examination of a system or a particular quality system component, issue, project, activity, or complaint.

Goal: Understand the detailed workings/failings of part of an organization or process, usually to answer a specific question.

Who conducts: Can be done by internal or external GMP auditors, including those from a contract giver or consulting group.

Comments: A focused audit is usually initiated because of a concern or issue that the audit sponsor has with the auditee. For example, a firm may conduct a focused audit on a vendor who has been having problems providing components that consistently meet the specifications.

Other questions that audit sponsors might have that would prompt a focused audit include the following:

- What is the auditee doing to resolve a problem?
- What have they learned from past problems?
- Is a department satisfactorily complying with a new corporate mandate (e.g., on employee training)?

Preparatory

Description: An examination of systems, practices, data, equipment, and facilities prior to an inspection by a regulatory agency, certification body, or other group. An emphasis is placed on those areas most likely to be inspected.

Goal: Identify weaknesses that can be corrected *before* the regulatory agency or auditing body arrives and prepare personnel for the upcoming audit.

Who conducts: Typically, internal GMP auditors or consultants.

Comments: With more firms moving toward ISO-9000 certification and the U.S. FDA's inspection activities being driven by their Pre-Approval Inspection Program, preparatory inspections are becoming more critical. Sometimes called "mock" inspections, auditors need to have a thorough understanding of the current certification and/ or regulatory expectations as they examine different areas.

Preparatory audits are also done in advance of corporate or customer visits.

One critical factor of preparatory audits is that they be done far enough in advance of the expected regulatory inspection that corrective actions can be planned and put in place before the agency or certifying body arrives. It is not unusual to conduct two or three preparatory audits with the later ones assuring the changes have been appropriately made.

Benchmarking

Description: Evaluating the specific practices, organization, and approaches used by other firms inside or outside of a company's own industry.

Goal: Understand what others are doing in order to improve one's own practices, products, services, and organization.

Who conducts: Quality assurance professionals or neutral, third-party consultants.

Comments: In the 1980s benchmarking was a quality management technique that had benchmarking teams traveling around the world looking for the best of the best. In the 1990s, benchmarking is still used, but it is more focused and goal oriented.

A dilemma in conducting benchmarking audits occurs when a firm wants to know how they compare with others, but does not want their competitors to know information about them. For this reason, a group of firms in the same industry will contract with third-party consultants who visit each firm in the group. Their report presents the data anonymously with each company being told only of their own identity. This type of anonymous reporting is not as important when firms are not competitors or when one firm is clearly recognized as having best practices.

Due Diligence

Description: An audit to determine risks that may exist in acquiring or merging with another organization.

Goal: Avoid harm to the reputation or financial status of the organization.

Who conducts: Auditor(s) from the purchasing firm or third-party consultants.

Comments: Due diligence audits are done by firms so they know exactly what they may be getting themselves (and their stockholders) into if they form a relationship with the auditee. These types of audits not only focus on GMP compliance issues but also are conducted on financial, environmental, and other aspects of the company. In each case, auditors with expertise in the particular fields must be involved.

Certification

Description: An audit conducted to determine if particular requirements have been met by the auditee.

Goal: Achieve certification.

Who conducts: Auditors from the certifying body or third-party auditors working for the certifying body.

Comments: One type of certification audit is done by a *registrar* as part of the ISO-9000 registration process. Auditors from a registrar typically have generic experience (i.e., they audit many different firms), not just with those who manufacture healthcare products. However, there is usually one audit team member knowledgeable in the industry sector involved in the audit (Quality Progress 1991). Registrars (and their auditors) need to maintain strict confidentiality and be free from conflicts of interest as they certify organizations (Weightman 1994).

If an organization is a finalist for a quality award, such as the Baldridge Award or California Quality Cup, firms typically are audited by a team working on behalf of the awards committee.

One other type of certification is being used more and more by the U.S. FDA. In this situation, firms that have been under an FDA consent decree (a court order requiring the drug or device firm to correct specific GMP practices) must hire consultants (approved in advance by the FDA) to certify that the firm has achieved compliance (The Gold Sheet 1996B).

Regulatory

Description: An examination by a regulatory agency (e.g., Canadian HPB, U.S. FDA) of the systems, process, facilities, equipment, management decisions, etc. involved in developing, producing, and distributing a drug product or medical device.

Goal: To determine compliance with current GMP regulations and expectations, as well as compliance with the agency-approved manufacturing process.

Who conducts: Agency personnel.

Comments: Regulatory audits (usually called *inspections*) are done with a frequency usually defined by law or regulations. It is this inspection activity that keeps most firms in (or attempting to be in) compliance.

In the U.S., FDA inspection programs are defined in the FDA's *Compliance Program Guidance Manual* (CPGM). Agency personnel's activities are defined in the *Investigation's Operations Manual* (IOM). Both of these documents are available to the public; see the Resources section for more information.

Conclusion

The type of audit is a concept that can help you better organize a GMP audit to accomplish a particular goal. Selecting the appropriate type of audit to use depends on the desired outcome of the audit and the organization to be audited. The types of audits presented in this chapter, while not a definitive list, cover most situations.

─────────────────────────────────────CHAPTER 6

Audit Approaches

Goals

- Define *audit approach.*
- Show the relationship between audit types and audit approaches.
- Identify when particular audit approaches may be used in conducting a specific type of audit.
- Provide examples of what to generally and specifically examine when auditing.

What Is an Audit Approach?

An audit approach is the set of methods and techniques used to collect information for a particular reason or related to a specific point of view. One approach can be used alone during an audit, or a combination of approaches can be used together to provide a more comprehensive view of the operation. Certain audit approaches will overlap to some degree with others.

Suggestions on which audit approaches to use are shown in Table 6.1. The decision on which one(s) to emphasize is based on the goal of the audit, past experiences with the auditee, problems/deviations, audit time available, and management comments.

Table 6.1. Suggestions on Using Audit Approaches

Type of Audit	Quality/ Mgmt	Preparatory General Insp	Preparatory PAI	Contract acceptor: Supplier	Contract acceptor: Contractor
Audit approach					
GMP QS	***	**	**	*	**
Documentation	na	**	*	*	*
Deviations	**	**	*	**	***
Validation/CC	**	**	***	**	***
Stability	**	**	*	na	*
Complaints	*	**	na	*	*
Data Quality	*	na	***	na	*
Equip/facilities	na	**	*	**	**
QC labs	na	**	*	**	***
Supplier reqs	na	na	*	***	na
Contractor reqs	na	na	na	na	***
Computer sys	*	*	*	na	na
Investigational product	#	#	***	na	na
Commercial product	na	**	na	na	na
General GMPs	**	**	*	**	**

(***) approach is highly recommended for this type of audit
(**) approach is suggested for this type of audt
(*) approach may be appropriate for this type of audit
(#) "Preparatory PAI" type of audit should be used
(na) approach is not a high priority for this type of audit

Examples of audit approaches include the following:

- GMP Quality System
- Documentation
- Deviations
- Validation, change control
- Equipment/facility
- Computer system
- Existing product
- Stability
- Complaints
- Supplier/contractor requirements
- Data quality and integrity
- QC laboratory
- New product development
- General GMP operations

Audit approaches make it easier to conduct an effective, methodical examination of significant issues. Points covered in the approaches presented below are based on the experiences of quality auditors, guidelines issued by regulatory bodies, and examples of what is currently important. Over time, this last point will change; auditors need to stay current with regulatory/inspectional trends, technology changes, and industry practice. Audit approaches are simply guides. They can easily be augmented and combined. Figure 6.1 gives an example for using a set of audit approaches when conducting an audit of a potential supplier.

In the audit approaches discussed below, you will find the following categories:

Question(s) to answer: The one or two fundamental questions that the auditors will attempt to answer as they examine the area, topic, etc.

Figure 6.1. Conducting an audit of a potential supplier of packaging materials.

Audit approaches to be used:

- Supplier requirements
- Validation/change control
- Deviations
- Documentation
- Facilities/equipment
- General GMPs

General things to look for: Typical, more global things that would be examined during the audit. In creating the audit checklist, the auditors would choose the general items that are most appropriate and then expand upon them with specific checklist items.

Specific examples: Suggestions that could be included in a checklist.

Audit Approaches: Quality System

Question to Answer

Is there an integrated system in place that keeps management informed about the firm's products, operations, quality, trends, and compliance activities?

General Things to Look For

- Formalization of the system (e.g., SOPs).
- Quality system components (Table 6.2).
- Functioning of the overall system.

- Functioning of the individual quality system components.
- Information feedback systems (e.g., follow-up and closure).
- Management knowledge and involvement.
- Quality measures and trending.

Note: Individual quality system components can also be examined in depth as needed using a focused type of audit.

Specific Examples

- There are paper records that support what has actually taken place.
- Management is informed of events; their actions are appropriate, justified, and documented.
- The facility operates in a state of control.
- All levels of the organization share ownership for producing quality products.

Table 6.2. Examples of Quality System Components

- Documentation
- Change control
- Material and product testing/release
- Calibration, preventative maintenance
- Validation
- Training
- Stability testing, monitoring
- Product quality reviews
- Notifications to management
- Complaint handling
- Problem reporting, investigation

Audit Approaches: Documentation System

Question to Answer

Is there an integrated process of developing, approving, controlling, distributing, and archiving functional documents (e.g., specifications, SOPs, protocols, lab methods) and completed records (e.g., logs, batch history records)?

General Things to Look For

- Formalization of the system (e.g., SOPs).
- People have the correct documents when they need them.
- Documents are controlled with ID and revision numbers.
- Documents have been reviewed and approved before use.
- Users do not have to make choices about which documents should be used.

Specific Examples

Functional Documents

- SOPs, methods, protocols, specifications, etc. are available where and when they are needed (i.e., at the "point of use").
- Documents have been reviewed and approved; training has been provided *before* the procedures, for example, are effective (Figure 6.2).

Figure 6.2. Documentation timeline.

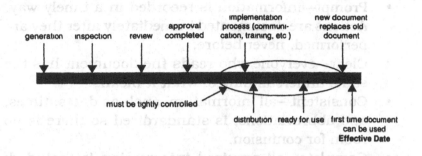

- People are performing the tasks according to the SOPs, protocols, and methods.
- Uncontrolled copies of documents are not used.
- Master folders/files of documents are arranged systematically and archived information is readily available.
- Documents are written clearly with the users in mind.
- Documents are reviewed periodically for their accuracy, relevance, and compliance to GMP expectations.
- Record retention period and archiving facilities are suitable.

Records

Properly prepared records have the following nine characteristics:

- Permanent—the information can't be changed, erased, or washed off.
- Legible—the information can be easily read.

- Accurate—calculations are correct and other information is recorded with care.
- Prompt—information is recorded in a timely way. Actions are documented immediately after they are performed, never before.
- Clear—everyone who reads the document has the same understanding of what it means.
- Consistent—all information, such as dates, times, and abbreviations, is standardized so there is no room for confusion.
- Complete—all required information is included; there are no unexplained "blanks."
- Direct—information is recorded immediately onto the proper form, lab notebook, or computer system.
- Truthful—all information included in the document is, to the writer's knowledge and ability, what really happened.

Audit Approaches: Deviations

Question to Answer

Are problems, deviations, out-of-specification (OOS) results, etc. properly documented, investigated, and corrected as appropriate in a reasonable amount of time?

General Things to Look For

- Formalization of the system (e.g., SOPs).
- Tracking of open cases.
- Resolution of the problem occurs *before* the lot is released.
- Quality unit involvement.

- A timeframe from the point of observation to reso-
.lution is defined and followed.
- Investigations are extended to other lots, products,
etc. as needed.
- Knowledge from deviations is applied throughout
the organization.

Specific Examples

- The relevant SOPs include specific requirements
about the investigation process, reports required,
follow-up, and timeframes.
- People have been trained in the SOPs and related
investigational techniques.
- There is a log or tracking system for deviations.
- If an investigation is *not* extended to other lots or
products, there is a clear, valid reason.
- Root causes of problems are determined using a
rigorous, rational investigational process.
- Problems do not recur.
- Management is kept informed of the number of
problems and their GMP/quality/business implications.

Audit Approaches: Validation and Change Control

Questions to Answer

Is there documented evidence to show that processes,
equipment, lab methods, cleaning practices, computer
systems, etc. do what they are intended to do on a consistent basis? Are changes appropriately evaluated with
thought given to compliance, validation, and training?

General Things to Look For

- Formalization of the system (i.e., SOPs).
- An approved validation master plan.
- A close tie exists between validation and change control.
- Proposed changes are considered before they are put into place.

Specific Examples

Validation

- Compendial methods used (e.g., from the U. S. Pharmacopoeia [USP]) have been shown to work in the local lab.
- Non-Compendial lab methods used for testing (e.g., release, critical in-process tests, stability) have been validated.
- Validation protocols are approved by technical personnel and the quality unit *before* being used.
- Critical processing/quality parameters that have been identified as needing to be controlled and validated.
- Validation reports that include a conclusion as to whether the system/process is or is not validated.
- There is a systematic way of conducting installation qualification (IQ), operation qualification (OQ), and performance qualification (PQ) for utilities, new equipment, and processes.
- Process validation has been conducted on all products.

- Critical processing parameters are validated and are consistent with the in-process specifications, SOPs/work instructions, and machine capabilities.
- Cleaning practices and materials have been validated for all products or a rational set of "worst-case" products.
- Computer systems and process logic controllers have been validated and include the appropriate level of security, back-up, and disaster recovery.
- All prospective validation studies have been done with a minimum of three lots in a row, without any changes between runs.
- Any retrospective validation has included a large series of runs that have not been specially selected; "failures" comprise less than 10% of the series.
- "Vision systems" in packaging have been validated.

Change Control

- There is a definition of "change".
- If there are different levels of "change" (i.e., major, minor), there is a rational, consistent way of discriminating between them.
- Changes in process, specification, methods, protocol, etc. are approved *before* they are implemented.
- The impact of the change *vis-à-vis* validation, customer requirements, and the approved drug's application (e.g., NDS, ANDS, NDA, or ANDA) have been considered.
- For a given SOP, method, product, or process, there is a log of changes made.
- There are examples of situations when changes have *not* been approved.

Audit Approaches: Stability Testing and Monitoring

Question to Answer

Is there a systematic and formalized way of initiating, generating, evaluating, and using stability data?

General Things to Look For

- Formalization of the system (i.e., SOPs).
- How products, package sizes, lots get selected for inclusion in the stability program.
- A definition of "special" lots needing stability testing and a description of how these lots are captured.
- Calibration and monitoring of the stability (environmental) chambers.
- Stability testing methods are valid indicators of stability.

Specific Examples

- The correct (expected) number and types of samples are actually on stability testing.
- Samples are being tested within a defined time period of their due dates.
- The appropriate parameters are being tested (e.g., potency, degradation products).
- The test methods have been validated as stability indicating.
- For a given product, there are no "holes" or missing data.
- "Special" lots, e.g., reprocessed, changed, deviations, are on stability testing.
- Stability test results support expiry dating.

- Stability test program supports storage recommen-.dations.
- Stability testing chambers are properly maintained, monitored, and alarmed.
- SOPs define what is to happen in the event of a stability failure.
- If a stability failure exists, it is handled as defined by the SOPs.

Audit Approaches: Complaints (Product)

Question to Answer

Is there a systematic way of receiving, documenting, investigating, and resolving complaints in a timely manner?

General Things to Look For

- Formalization of the system (e.g., SOPs).
- How complaints are processed and investigated.
- Rationales exist for not investigating certain types of complaints.
- Complaints are investigated and completed in a reasonable amount of time.
- Communication with complainant is professional, courteous, and informative.
- Trends of complaint data are reviewed by management.

Specific Examples

- Complaint files exist and are readily available.
- If a complaint is not investigated, a valid rationale is given.

- Complaint data can be sorted and viewed different ways (e.g., time, product, location, etc.).
- There is an SOP that describes when and how a complaint is to be investigated.
- Qualified personnel investigate complaints.
- If necessary, a complaint investigation extends to other lots, products.
- Management is informed about significant complaints and complaint trends.
- Criteria exist on when to notify regulatory agencies (HPB, FDA, etc.) about significant situations.
- Employees are informed about comments on the products they make.
- Complaint information is used to improve products; actions are linked to approved changes.

Audit Approaches: Data Quality and Integrity

Question to Answer

Can the data generated during development, testing, and other GMP activities be trusted as a complete, accurate, and truthful picture of what really happened?

General Things to Look For

- SOPs for data collection and recordkeeping.
- Data correctness and corroboration.
- Data "reasonableness."
- Conclusions based on all data, not just selected data points.

Specific Examples

- Lab notebooks and documents are clear with a reasonable number of corrections.
- Information attached to or included with documents is done in such a way to minimize the chance for later changes.
- Data are complete; missing information is not casually dismissed.
- Records and reports are readily available.
- Records, logs, notebooks, etc. look "real" and used.
- The original data that are the basis for a summary report (e.g., graph or table) that was submitted to a regulatory agency can be easily obtained; the data support all conclusions made.

Audit Approaches: Laboratories (Microbiological, Analytical, and Developmental)

Question to Answer

Do the laboratories produce scientifically defendable results that can be relied upon for making decisions about raw materials and in-process and finished products?

General Things to Look For

- Definition of roles (organization chart).
- Methods are qualified and/or validated.
- Samples, standards, reagents are identified, controlled, and properly stored.

- Instruments are properly calibrated and maintained.
- Unusual situations or results (OOS) are documented and investigated.
- Change control is practiced.
- People are qualified and properly trained.
- Data are properly collected, recorded, and retained.

Specific Examples

- SOPs exist for equipment cleaning, maintenance, and calibration; SOPs are being followed.
- Reference standards are properly labeled, dated, and stored.
- Reagents are labeled with use/expiration date; any preparation has been recorded.
- Validation records are readily available.
- In microbiology labs, autoclaves operate within the correct temperature range.
- In microbiology labs, identification of isolates is done with consideration of the product and its intended use.

Audit Approaches: Vendor (Supplier) Requirements

Question to Answer

Can/do vendors produce products that consistently meet the specifications of the customer and comply with all requirements and applicable regulations?

General Things to Look For

- Formalization of how vendors are qualified (e.g., SOPs).
- Defined, agreed-upon requirements and responsibilities.
- Complete histories of products made.
- A system for keeping vendors/suppliers up-to-date with specifications, changes, drawings, etc.
- Problems/deviations are investigated and resolved.
- Conformance to GMP principles.

Specific Examples

All Vendors, Contractors

- Information and products are maintained to minimize mix-ups and diversion.
- Identity control is maintained.
- Specifications, drawings, requirements used by supplier are the correct, current versions.
- Processes are validated or are running consistently with few problems.
- Data generated (e.g., Certificate of Analysis) is reliable and is periodically confirmed.
- Identification and assignment of lot numbers is done rationally and in keeping with requirements.
- Cross-contamination is prevented by proper facilities, materials/personnel flow, and HVAC systems.
- Vendor/supplier notifies firm of changes to process *before* change occurs.

Note: Use other audit approaches as appropriate for the supplier's activities.

Active Pharmaceutical Intermediates (i.e., Bulk Drug Substances)

- The manufacturer operates in accordance with GMP.
- The supplier has an independent, functioning quality unit.
- The API is made using a validated process.
- Equipment cleaning is adequate and validated.
- Type of water used is appropriate for the product in which the API will be used.
- The identity and release status of all materials, solvents, etc. used can be easily observed.
- The API supplier formally defines "batch" or "lot"; manufacturing histories can be traced.
- Written procedures describe the handling of mother liquors and reworked material.
- Shipping containers for final APIs have been cleaned using a validated process.
- Personnel are qualified and adequately trained for their tasks.

Raw Materials, Components

- The original source of material is known.
- The supplier formally defines "batch" or "lot"; manufacturing histories can be traced.
- The supplier uses SOPs, specifications, and some form of manufacturing records.
- Raw materials used by the supplier are of suitable quality.
- The supplier has an independent, functioning quality unit.
- Complaints and problems are investigated.
- Personnel are qualified and adequately trained for their tasks.
- Supplier monitors/trends the quality of its product.

Labels and Labeling

- The supplier has adequate control/security over printing plates and finished printed material.
- The supplier has controls in place to ensure only currently approved versions are used.
- Rejects and waste are properly destroyed.
- Paper stock and inks are properly stored; stock is rotated.
- The printing process used minimizes opportunity for mix-ups.
- In-process controls for color matching, print quality, etc. are used.
- Supplier advises firm in advance of any changes.
- Cleanliness of facility is appropriate.
- The supplier has an independent, functioning quality unit.

Audit Approaches: Facilities and Equipment

Questions to Answer

Is the facility capable of consistently producing a product that is fit for use, safe, identified, of the proper strength, pure, and of high quality? Do the facility and its equipment meet current GMP/regulatory standards?

General Things to Look For

- Procedures for maintenance, sanitation, pest control, utility operation, and equipment operation, set-up, and cleaning.
- General appearance.

- Housekeeping and cleanliness.
- Material and people flow through the process/facility.
- Potential for cross-contamination.
- Suitability of the facility, equipment, and operations to the product(s) or activities being performed.
- Preventative maintenance is defined and occurs according to a rational schedule.
- Calibration is defined and occurs according to a rational schedule.

Specific Examples

- As-built drawings are available for the facility and utilities.
- Changes to utilities, water systems, etc. go through the change control process.
- Surfaces are constructed of cleanable materials.
- Lists of approved cleaning materials are available; only these materials are available for use.
- Manufacturing records exist for preparation of cleaning solutions or sanitization agents; expiration dates.
- Pest control practices are successful.
- Cleaning practices prevent cross-contamination and generation of particles.
- Drainlines have an airbreak before entering drains.
- Hoses are hung for proper draining.
- Restrooms are maintained with appropriate soaps, towels, dryers, etc.
- Environmental monitoring takes place as needed (e.g., particulate and microbial monitoring).
- Preventative maintenance and calibration activities are documented.
- Calibration standards are traceable to official standards.

Note: Other audit approaches, such as validation and

change control, deviations, computer system, or general GMP operation, can be used as appropriate when examining facilities and equipment.

Audit Approaches: Computer Systems

Question to Answer

Is the computer system secured, validated, supported, backed up, and maintained so it will consistently perform according to its requirements?

General Things to Look For

- A rational approach toward validation of the computer system, considering hardware, software, networking, etc.
- Definition of the system (hardware and software).
- Approved validation protocols.
- The quality unit (QA/QC) is involved in review and approval of all protocols and reports.
- Rational approach toward validating/verifying applications (e.g., Excel) used for GMP activities.
- Evidence that the validation was performed according to the protocol.
- Conclusions of the validation (e.g., was it successful or not?).
- Implementation of change control.
- Disaster recovery plans.

Specific Examples

- Custom-written applications were prepared to an approved requirements document.

- Good programming standards (e.g., ISO 9000-3) were used by the software writers.
- Only authorized people have access to the system and terminals.
- If electronic signatures are being used, regulatory requirements are being met.
- Back-ups occur routinely.
- Data cannot be modified once entered; there is a defined process for making corrections and amendments.
- Problem logs are kept.
- If the system cannot be used, validated alternatives are available.

Audit Approaches: Commercialized Product

Question to Answer

Are commercialized (i.e., marketed) products made according to the agency-approved submission so they meet regulatory requirements *and* current GMP expectations?

General Things to Look For

- Compliance to agency-approved submission (e.g., NDS, NDA).
- Validation of the process being used.
- Communication with the agency on changes.
- Periodic quality reviews.
- Trending of product failures, rejections, withdrawals, and recalls.

Specific Examples

- Internal files that show exactly what has been registered with the agencies.
- Products have been properly validated (prospectively or retrospectively [for "old" products]).
- Analytical methods related to the product have been validated.
- Equipment cleaning methods have been validated.
- For U.S. products, an annual product quality review is prepared with conclusions, QA approval, and follow-ups as required.

Audit Approaches: Investigational Product

Questions to Answer

Are there raw data to support all claims and submissions made to the agency (e.g., HPB, FDA)? Are the facilities, equipment, and personnel capable of manufacturing the product according to the submission?

General Things to Look For

- Overall quality of the raw data.
- Rationales for important decisions made during the development process.
- Identification of critical processing parameters and the specifications.
- General GMP compliance.

Specific Examples

- Raw data are readily available that support a graph or summary table in the submission.
- A development report (though not required by the GMPs) is available that describes the development process, identifies critical processing parameters, and gives rationales for important decisions.
- Installation qualification (IQ) and operation qualification (OQ) have been performed on development and production facilities, utilities, and equipment.
- Critical processing equipment (e.g., autoclaves, WFI systems) has been performance qualified (PQ).
- If the process has not been validated, there are protocols that have been approved.
- Analytical methods related to the product have been validated.
- Equipment cleaning has been validated.

Audit Approaches: General GMP Operations

Question to Answer

Does the drug manufacturing firm consistently operate in a way that meets all GMP requirements and expectations?

General Things to Look For

- Definition and functioning of the quality unit.
- Integration of GMPs into the core business.
- Application of GMP requirements/expectations into each department.

- All levels of the organization work together to produce products that meet or exceed GMP requirements.

Specific Examples

Purchasing

- Works with operations and quality unit to approve suppliers from a quality/GMP point of view.
- Works with operations and quality unit to resolve quality problems with suppliers.
- Facilitates communication with suppliers.

Warehousing/Receiving

- Incoming materials/components are properly labeled according to identity, lot, and status.
- Physical/electronic systems exist for segregating materials according to status.
- Physical/electronic systems exist for quarantining materials.
- A first-in/first-out (FIFO) method of inventory control is used.
- Incoming materials are properly sampled, resealed, and stored.

Dispensing

- Dispensing rooms work on one lot/product at a time.
- Area and tools are properly cleaned.
- Scales and dispensing equipment are maintained and calibrated.
- Identities and amounts are verified/witnessed.
- Product and lot identities are maintained.

Manufacturing

- Facilities, equipment, processes have been installation qualified (IQ'd), operation qualified (OQ'd), and performance qualified (PQ'd).
- All equipment and materials are handled to maintain the safety, identity, strength, purity, and quality of the product.
- Activities are concurrently recorded with the action.
- In-process sampling is done as defined; in-process testing is comparable to that done by Quality Control (QC) lab.
- Cleaning is done using validated techniques.
- Cleaning agents/materials are defined; only these are found in the facility.
- Process control (i.e., statistical process control [SPC]) is used for tablet weight or fill weight; results are used to control the process.
- Environmental monitoring is done as appropriate.
- Appropriate levels of particulate and microbiological controls are used.
- Lots are subdivided to make problem control and resolution easier.

Aseptic/Parenteral Manufacturing

- Personnel use proper aseptic technique.
- Facilities, practices, and equipment are selected and implemented so as to minimize bioburden and particulates.
- Integrity testing on filters is used; alert and action limits are in place.
- Filter selection is rationally made, based on the type of product, its pH, temperature; filter media used do not have adhesives that will leach out under conditions of use.

- The firm has, if needed, documentation/certification from the filter manufacturer to support deviating from published filtration parameters.
- The firm tracks the identity of multiuse filters; actual number of uses does not exceed manufacturer's recommendations.
- There are time limits established for filtration; these limits are not exceeded.
- All personnel who enter the aseptic area gown properly, according to a defined SOP.
- Aseptic operators and others who enter the aseptic area are monitored for possible microbiological shedding and contamination.

Packaging and Labeling

- Labels and labeling are tightly controlled in terms of storage and release-to-areas.
- Expiration dates, lot numbers, etc. are carefully monitored and verified before labeling begins.
- Packaging line areas are operated and movement between lines is minimized so as to prevent mix-ups.
- Line clearance methods that include particular places to inspect have been defined.

Utilities and High Purity Water Systems

- Systems have been validated.
- Systems are monitored.
- Personnel have been trained in collecting samples, and they perform sampling correctly.
- Action limits are specified, along with responses to be taken.
- Variations in quality of incoming materials (e.g., source water) do not affect quality of final product.

Maintenance

- Facilities are properly maintained.
- Maintenance personnel who enter restricted areas are adequately trained, monitored, and dressed.
- Tools, equipment, instruments are cleaned and maintained so as not to contaminate or cross-contaminate facilities.
- Sanitation, pest control activities, and preventative maintenance are defined by SOPs and schedules.
- Areas are notified well before any maintenance activities take place.
- Maintenance and change control are tightly connected.

Training

- SOPs exist, are available, and are used; people have been trained in them.
- Personnel have the documented qualifications and training they need to perform their jobs.
- Training is done on a periodic basis.
- There is evidence to show that the training has been effective.
- Training is performed by qualified personnel.

Conclusion

Audit approaches are guides that GMP auditors can use as they plan particular audits. Auditors can shape an approach to better match the specific needs of the audit and their own auditing style. Audit approaches should be periodically reviewed so they stay current with the ever-changing regulatory, technology, industry, and quality issues.

Auditor Qualifications and Skills

Goals

- Identify regulatory requirements on GMP auditor qualifications.
- Identify ISO expectations for quality auditors.
- Discuss three sets of skills a GMP auditor needs to have.
- Discuss training options available to auditors.
- Describe auditor certification, what it means, and what it doesn't mean.
- Describe what makes a good auditor.

No one is born a GMP auditor. As we will see, a successful auditor has abilities and skills in three main areas. Those skills are developed with education, experience, and awareness.

What Is Required of a GMP Auditor?

As shown in Table 7.1, the requirements for GMP and quality auditors are rather vague, presenting both a challenge and opportunity to firms. The opportunity is being able to define what *qualified* means in your organization. The challenge comes in meeting those qualifications.

The ISO has issued its *Guidelines for Auditing Quality Systems* that includes criteria for quality systems auditors (ANSI/ISO/ASQC 1994) that can be applied to GMP auditors as well. The Guideline states that "auditor candidates" must have particular education, training, experience, personal attributes, and management capabilities.

- Education. Quality auditor candidates need to have completed at least secondary levels of education. They must have "demonstrated competence in clearly and fluently expressing concepts and ideas orally and in writing in their officially recognized language."
- Training. "Training in the following areas should be regarded as particularly relevant:"
  — knowledge and understanding of the standards. . .
  — assessment techniques of examining, questioning, evaluating, and reporting. . .
  — planning, organizing, communicating, and directing [the audit].
- Experience. "Auditor candidates should have a minimum of four year's full-time appropriate practical workplace experience, not including training," at least two years of which should have been in quality assurance activities. . . .
- "Prior to assuming responsibility for performing audits as an auditor, the candidate should have. . . participat[ed] in a minimum of four audits for a total of at least 20 days, including documentation review, actual audit activities, and audit reporting."

Table 7.1. Summary of Requirements for GMP/Quality Auditors

Standard	Reference	Statement (emphasis added)
Canadian Drug GMPs	Section C.02.010, Interpretation 2.2	[For a raw material packaging component supplier certification program that includes an on-site audit] "an on-site audit . . . by a **qualified person . . ."**
	Section C.02.017, Interpretation 2.2 Section C.02.012, Interpretation 2.3	"The [internal] self-inspection team includes staff and others who are **suitably trained and qualified with GMP."**
European Community GMP Guidelines	9.2	"Self-inspections should be conducted in an independent and detailed way by designated **competent** person(s) from the company."
U.S. Drug cGMPs (21 CFR 210-211)	(Topic not covered)	
U.S. Medical Device Quality System Regulations (21 CFR 820)	820.25	"Each manufacturer shall have sufficient personnel with the necessary education, background, training, and experience to assure all activities . . . are correctly performed." *
ISO-9001 (ASNI/ASQC Q9001-1994)	4.17	"Internal quality audits . . . shall be carried out by personnel independent of those having direct responsibility for the activity being audited."

* In the Quality System Regulation's Preamble, the FDA referred to this section that refers to auditor qualifications (FDA 1996c).

- Personal attributes. "Auditor candidates should be open minded and mature; possess sound judgment, analytical skills, and tenacity; have the ability to perceive situations in a realistic way, to understand complex operations. . ."

The ISO standard continues (Annex A) by describing methods of evaluating auditor candidates, such as interviews, reviews of written work, role playing, observation during an actual audit, and structured testing.

Industry practice in the pharmaceutical and medical device industries can help clarify what *qualified* means. Most, if not all, GMP auditors have had a combination of training, coaching, and experience to be qualified, though not as stringently applied as the ISO requirements above. Sometimes this will involve participating in in-house training courses or attending outside workshops. Many auditors have received a significant amount of their training by working alongside experienced auditors.

During the mid-1990s, there were a number of GMP auditors who entered industry and consulting firms who learned auditing techniques while working for the U.S. FDA. Another trend has been auditors who have been certified as Quality Auditors (e.g., CQA designations) by professional associations. Your firm needs to establish its own criteria for what *qualified* means and then provide a path for hiring and/or developing GMP auditors to meet those criteria.

Three Areas of Competence for Auditors

In the pharmaceutical and medical device industries, GMP auditors need to have competencies in three particular areas. This model, which includes almost all of the points described in the ISO document, is shown in Figure 7.1.

Figure 7.1. Three areas of competence all GMP auditors need to have.

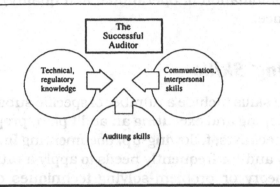

Technical and Regulatory Knowledge

To be skilled as a GMP auditor, you must understand what you will be seeing. While quality auditors (e.g., auditors for ISO 9000 registration bodies) don't need a great deal of experience in the specific industry they are examining, most GMP audits are different. GMP auditors need to know various ways of doing a task, such as conducting a stability assay, and the critical issues that the firm should address. This is especially important for complex utilities (e.g., Water-for-Injection installations) or processes (e.g., aseptic manufacturing).

Regulatory requirements and expectations are another set of knowledge needed by a GMP auditor. Although most auditors will use checklists or guides, a successful auditor must understand how agency inspectors are *currently* interpreting the regulations. He or she must be able to evaluate a particular practice and determine how it affects the safety, identity, strength, purity, quality, and fitness-for-use of the product. While some of this knowledge can come from formal education or training sessions, a major portion of it can only come from living with it for years. This is one reason GMP auditors are not born or hatched right out of school; GMP auditors need to have some combination of operational and quality/regulatory experience.

Auditing Skills

Auditing skills include a number of specific subskills, such as developing and executing an audit plan, preparing and using checklists, following-up, documenting findings, etc. Also, an auditor frequently needs to apply a rational sampling theory or problem-solving techniques during the audit.

Some auditing skills are developed over time and made sharper with practice. For example, being able to ask questions that are deceptively simple, yet cut to the heart of the issue. Being able to analyze a situation is also important for an auditor, separating elements that are associated with an issue from those that are causing it. Related to this is being able to see what is really important in a situation: seeing the forest AND the trees for what they are as well as the relationship they have to each other.

Communication and Interpersonal Skills

If you think of the technical and regulatory knowledge of the audit as the content and auditing skills as the process, communication and interpersonal skills are the lubricants. If the auditors communicate and interact well with everyone involved, the audit will be successful. If not, there will be friction, stress, and organizational trauma.

Almost every GMP auditor, as well as those who have been audited, has horror stories of visiting auditors who were unreasonable, petty, unjustifiably arrogant, or incredibly difficult. Here are some examples:

- . During an inspection of a labeling area, an auditor insisted that all waste containers be emptied on the floor so he could see if labels were being disposed of improperly.
- An auditee invited the corporate auditor (from another country) to spend part of the weekend visiting an archeological site, which was a national treasure. The response from the auditor, "Why do I want to spend my day off looking at piles of old rocks?"
- An auditor insisted on including several minor, isolated deficiencies on the final report because he said, "Somebody has to be the bad guy and I love being him."
- An auditor with a reputation for not sharing all of his findings during the verbal presentation; auditees were frequently surprised by an additional major finding listed in the final report.

Auditors have an important role beyond that of simply auditing: they represent the sponsoring firm. Some characteristics of good interpersonal and communication skills in an auditor include the following:

- A true respect for people.
- Sensitivity to cultural differences.
- Active listening (being sure you understand what is being said).
- Spending more time listening than talking (a three-to-one ratio).
- Oral comments that are not rambling.
- Written comments that are succinct and really say what was intended.
- Comments that are focused on behaviors and objects, not on people.

- Comments and actions that are viewed as impartial and objective.
- A decisiveness in decision making as well as an ability to say "I don't know" when appropriate.
- Motivating people to do things when they are reluctant or resistant.

Ethics and Integrity

Ethics, integrity, and overall professional behavior on the part of the auditor are qualities that cannot just be assumed. If the auditee has a sense that the auditor is anything less than totally honest and impartial, the audit will not be successful. Sometimes the audit sponsor undermines the professionalism of the audit by having a secret agenda for the auditors, such as finding reasons to end an agreement, cancel a contract, or identify people to recruit. This places the audit team in a difficult situation. Other aspects of ethics and integrity that auditors need to have include those listed below:

- Keeping confidential materials confidential, not only by not talking about things inappropriately, but, for example, by not reading proprietary documents on crowded planes.
- Freedom from conflicting interests or full disclosure of them.
- Independence from forces that could cause undue pressure or influence.
- Being straightforward about issues and implications.
- Not causing intentional surprises or embarrassment to those being audited.

Frequently, second- and third-party auditors sign confidentiality agreements or non-disclosure pacts with the auditees. Maintaining high levels of personal and professional integrity will supersede the requirements of these legal documents.

If You Are Not the "Perfect" Auditor . . .

There are very few GMP auditors who have high levels of technical/regulatory knowledge, auditing skills, and interpersonal/communication skills. In real life, auditors can compensate for lower expertise in one area with stronger skills in another.

Audit Teams: Sharing and Building Expertise of Auditors

A benefit of using teams of auditors is that they can share expertise and help to support the learning of their colleagues. In using teams of auditors where there are varying levels of expertise, the team leader must be very knowledgeable in both auditing and at least some of the technical issues.

Becoming a Qualified Auditor

Your organization needs to define what it considers a "qualified auditor" to be. One approach is shown in Table 7.2.

What to Expect as an Auditor

Along with the "glamour" of auditing plastic molding firms during a New England snowstorm or a bulk chemical facility in Brazil in December, there are other situations that an auditor needs to be prepared for.

Table 7.2. Example of a "Ladder" and Qualifications for GMP Auditors

Position	Requirements to be in this position	Ways to demonstrate competence	What you should be able to do in this position
Auditor-in-Training	Academic training (secondary school graduate). Technical expertise in at least one relevant area. Experience in interpreting or applying GMP regulations. Communication skills. Have a sensitivity to ethics and integrity issues.	Work history or résumé. Interviews with technical/quality experts. Written documents (e.g., reports, other written commu-nication).	Participate in an audit as an "auditor-in-training." Assist in preparing checklists, developing an audit plan and action plan.
Auditor	Have participated in at least three audits of two different types. Formal training (in-house or external) in GMP and regulations. Have a strong sensitivity and understanding of ethics and integrity issues.	Evaluation by lead auditor and other auditors of the candidate's technical and regulatory know-ledge, interpersonal skills, auditing skills. Successfully played major role in develop-ing audit plan and report.	Contribute to the success of an audit. Work with minimal supervision. Evaluate deficiencies consistent with views of other technical and regulatory experts. Successfully execute an audit plan. Communicate effectively.
Lead Auditor	Participate as an auditor in at least 5 different audits, including supplier, preparatory, quality/management audits.	Evaluation by other lead auditors of candidate's auditing skills, organizational abilities, commu-nication skills.	Manage an audit without supervision. Critique audit plans. Critique audit check-lists. Provide advice to

Continued on next page.

Continued from previous page.

Follow-up and close out an audit under supervision of a Lead Auditor. Have a well developed sense of personal and professional ethics and integrity.	Display leadership during audit; perform with a minimum of supervision. Optional: certification by a professional organization.	auditors. Close out the audit to the satisfaction of the audit sponsor.

Not everyone will be happy to see you. For most auditees, audit teams are an unwanted distraction from what they would rather be doing.

It will be difficult to satisfy both the audit sponsor and the auditee. Your job will be to assess reality and present a report to the sponsor; the auditee typically wants to be shown as excelling in all they do.

A variety of surprises may cause deviation from the audit plan. The plan is developed to accomplish the goal of the audit; however, sometimes circumstances force a change.

Conclusion

What makes a successful GMP auditor? In summary, the successful auditor possesses the following characteristics:

- A practical understanding of the area to be audited.
- Understanding of the requirements and why they are important.
- Discipline and curiosity.
- Communicates effectively in writing and speech.
- A respect for others and a desire to learn from them.
- Belief in the importance of what he or she is doing.
- Honesty and integrity.

CHAPTER 8

Phase I: Preparation

Goals

- Identify the tasks that need to be accomplished in preparing for an audit.
- Describe the relationship between the goal of the audit and the audit type and approach(es) that are selected.
- Discuss an audit plan and what it includes.
- Discuss an audit action plan and what it includes.
- Describe checklists and how they are developed.

Preparing for the audit involves arranging the details of the audit, developing the audit plan, identifying what will be examined, and communicating with the auditee. Preparing for the first audit of a particular type and area (e.g., a GMP operations audit of a warehouse, a focused audit on deviations) is the most time-consuming, as you will need to develop checklists and other auditing tools. Once they are created, they can then be repeatedly used with only slight revisions and updates.

The preparation phase of an audit takes time to do properly, but it does have a payoff: the time spent auditing will be more productive, with all involved having realistic expectations that will be met. Preparation happens in a planned, structured way. It happens far in advance of conducting the audit. It is not done in the car or on the flight to the auditee's facility.

Who Is Involved

The team leader and audit team members are primarily involved in Phase I. The auditee and the sponsor's management are involved to a lesser extent.

When It Is Done

Scheduling an audit is sometimes done months in advance. When going through the preparation for the first time, you should begin creating the checklists and other audit tools several months before the audit is to commence. If checklist development is not needed, you should begin to review the materials and correspond with the auditee 4–6 weeks before the scheduled date of the audit.

Key Tasks

Activities that are accomplished during the preparation phase include the following:

- Define the goal and scope of the audit.*
- Identify areas and/or products to be audited.*
- Select the type of audit.*
- Define the audit approach to be used.*

- Identify the team leader ("lead auditor") and mem-
  bers; define roles and responsibilities.*
- Define the standards (expectations) and rating sys-
  tem to be used.*
- Identify a date and tentative audit schedule.
- Identify key auditee contacts.*
- Define audit report format and distribution.**
- Develop checklists; share checklists with manage-
  ment and quality unit.*
- Confirm dates, schedules with auditee.
- Make travel arrangements as needed.
- Share goals, checklists, expectations, etc. with
  auditee; identify additional goals and requests.*
- Request background documents from auditee.*
- Review historical (file) documents and previous
  audit reports.*
- Develop initial process flow diagrams.
- Start collecting working papers.*
- Prepare detailed action plan with schedule, priori-
  ties, team member assignments.*

* Discussed in more detail below.
** Discussed in Chapter 12.

The Audit Plan

The first eight items in the key task list make up the *Audit Plan* (Figure 8.1 A, B), a document that is provided to the firm's (or sponsor's) management and to the auditee. Information from the audit plan is also used in the final report.

Audit Goal

The goal is the purpose of the audit. Examples of goals include the following:

- To prepare for an upcoming regulatory inspection,
- To qualify a potential supplier, or
- To evaluate how the GMPs have been applied in a given area.

The audit goal is used in selecting the type of audit; the conclusions made in the audit report are also linked to the goal.

Areas or Products to Be Audited

Areas or products to be audited are the organizational areas, such as product design, liquids manufacturing department, or the microbiology lab. Listing the areas to be audited describes *where* the audit will take place or products that will be covered. As with the audit scope, the area or products to be audited need to be reasonable for the time and people available. As you plan the audit, be sure you know the physical locations, as some operations may be spread around a large campus or at different locations; you may need to allow for travel time between these sites.

Audit Scope

The scope defines *what* will be looked at during the audit. For instance, it may be the development of an investigational new drug product; the production, testing, and release of an active ingredient purchased from a vendor; or the final products warehouse. The scope puts boundaries on what will be examined; it also needs to be reasonable for the time and people available for the audit.

Figure 8.1A. Example of an audit plan (first page).

Audit Plan

For Training of Technical Ops Personnel
October 1997

Goals
1. Determine compliance with SOP 748/2 (Training of Tech Ops Employees) and CP 15/1 (Corporate Policy–Training).
2. Identify barriers in complying with SOP and CP.
3. Recommend solutions for improvement.

Scope
Technical operations training program.

Areas to Be Audited
- Manufacturing and packaging at Corporate Technology Center
- Engineering at Corporate Technology Center
- Warehouse and Distribution Center, Corrls St.
- Quality Assurance at Corporate Technology Center
- Materials Management at Corporate Technology Center

Approach
General GMP Operations, specifically training.

Type of Audit
Focused internal audit.

Audit Team

Lead auditor: Frank Roussel

Team members: Viveca White, Janice Newmar, Ida Peel, and Julie Brown

---

Figure 8.1B. Example of an audit plan (second page).

Auditee Contacts
Area training coordinators.

Audit Standards to Be Used
Five categories (defined in SOP 943/1):
1. Outstanding
2. Satisfactory/adequate
3. Minor deficiency
4. Major deficiency
5. Critical deficiency

Proposed Audit Schedule
Oct. 8, 8:30–16:00—Corp. Tech. Center: Manufacturing and Packaging
Oct. 9, 8:30–16:00—Corp. Tech. Center: QA, Engineering, Materials Management
Oct. 15, 8:30–12:00—Corris Street: Warehouse and Distribution

General Information
    An audit checklist will be used to guide the auditors; copies to be provided to auditee areas by Sept. 10.
    Anonymous questionnaires will be given to selected associates to get their reactions to current training activities and identify barriers to effective training.
    At the beginning of the audit in each area, a "kickoff" will be held to review the goals, general audit process, and answer any questions.
    At the end of the audit in each area, a wrap-up session with auditee contact and manager will be held to present initial findings, verify perceptions, and discuss recommendations.

Key Auditee Contacts

Early in the process, you should identify a contact who is the "gatekeeper" to whom you can funnel information. This also might be the person who will accompany the auditors during the actual audit.

Audit Team

The audit team includes one to three people (or sometimes one or two more) who will carry out the audit. Members should have particular expertise they can apply as they audit. The lead auditor must have previous experience in auditing; he or she is responsible for preparing the final audit report. (Table 8.1 lists typical responsibilities.)

While it might be tempting to try to get 5–7 people involved in an audit team, it is not really practical. The auditee will need to have a contact for each group that goes out into an area, which will cause an even greater strain on their resources. It will also be difficult to coordinate the team's efforts, both during the audit and when the audit report is being prepared.

Table 8.1. Responsibilities of the Audit Team

Team Leader	Team Member
• Lead, facilitate team • Be primary contact, spokesperson • Preparation of audit report • Collection, retention of working papers • Communicate wth sponsor management • Be a member of the team	• Apply expertise during audit • Assist with preparation • Document findings in an organized way • Working with auditees in a spirit of improvement • Perform with integrity and the highest ethical standards • Work with leader to accomplish audit plan, action plan in timely way • Communicate with leader

When selecting team members, find those who have expertise in several areas—they will add the most value to your team. Try to limit the number of new auditors to only one per audit.

Standards

Standards are the numbers or words used to classify the importance of a particular audit observation. The most basic standard is binary, that is, it includes only two possible responses, such as "yes" or "no." These answers indicate compliance or noncompliance with the requirement (Figure 8.2). The compliance or noncompliance standard is similar to what the U.S. FDA does during its inspections: Situations of noncompliance are recorded on a special form (FDA Form 483) as observations.

Another type of standard that is slightly more complex has five categories. In the scheme below, there are three types of deficiencies and two types of compliance situations.

1. **Critical deficiency.** The deficient situation(s) observed could significantly affect the safety, identity, strength, purity, and quality (SISPQ) of the product, resulting in a recall, regulatory action, or harm to the patient. Examples of critical deficiencies include mislabeling, potency above or below label-claim limits, and microbial contamination.

2. **Major deficiency.** The deficient situation(s) observed could affect the SISPQ of the product and would likely be cited during a regulatory inspection. Examples of major deficiencies include a cleaning procedure not validated, unauthorized process deviations, and undocumented preparation of lab reagents.

Figure 8.2. Example from a binary (yes/no) standard.

---

84. Is there an SOP describing the receipt of raw materials?
    ❑ YES   ❑ NO

85. Is there an SOP that describes QC sampling of raw materials?
    ❑ YES   ❑ NO

---

3. **Minor deficiency.** The deficiencies observed are not likely to affect the SISPQ of the product, but, nonetheless, do not meet current GMP (or organizational) expectations. Examples of minor deficiencies include recordkeeping errors, overdue SOP reviews, unauthorized process deviations.

4. **Satisfactory/adequate.** The requirement is met; the situation is not deficient in any way.

5. **Outstanding.** The requirement has been more than met in a way that displays excellence.

The Critical/Major/Minor standard is similar to that used by the British drug inspectors from the British MCA. Other options for standards include numbers for noncompliance situations; when totaled, the higher the number, the more noncompliant the situation. For example:

- 3—Critical deficiency
- 2—Major deficiency
- 1—Minor deficiency
- 0—Satisfactory/adequate

While not a standard *per se*, it is useful to note items that were not applicable in a situation or were not examined. A simple way to do this is to have "NA. Does not apply in this situation" available as an option.

What type of standard should you use? Many drug/device firms are using three to five categories for their standards. In making the decision on what to do, think about the following:

- Keep it simple.
- Will past audit results be of comparative value with your new standards?
- Will the standards be useful for several years to come so you won't need to change again in the near future?
- Will GMP auditors, management, and staff find the standards useful?
- Will auditees accept the standards and their definitions?
- Keep it simple.

As will be discussed in the following chapter, the number of deficiencies seen, as well as their significance, can be used in categorizing them.

Checklists

Checklists identify what is to be examined during an audit and define the desired level of performance. They guide the audit team as an audit is performed. Checklists also standardize audits allowing more consistent, comparable data to be collected over time. This is especially useful if management wants to compare the results, for example, of different teams who are auditing vendors.

A checklist is developed for each different area/product that is being audited. For example, in the audit plan shown in Figure 8.1, auditors would be provided with a set of checklists for each of the areas to be audited.

Drawbacks of Checklists

Some auditors resist using checklists because they feel that checklists intrude into the audit, reduce their craft to a mechanical process, and prevent spontaneous examination of areas that could be important. In some ways this is true: to have consistency over time and between people, there *are* tradeoffs that need to be accepted. In other ways, however, checklists are only guides; if the auditor thinks that a deviation is warranted, he or she should be able to make that professional judgment.

Another difficulty found with some checklists is that they are not designed according to how auditors work. For example, having a list of questions and standards with no room for the auditor to make comments, list documents examined, etc. makes it difficult for the auditor. One firm, in trying to more fully define their compliance auditing program, developed a 400+ page audit checklist for auditing their aseptic manufacturing operations. The resulting document, while elegant in concept, was extremely difficult to use. Auditors who tried using this spent more time turning pages than actually auditing. The project was abandoned.

Preparing Checklists

Checklists are developed using a variety of information sources and experiences, some of which are listed below. (Additional, specific references are found in the Reference section.)

- GMP Regulations (Canadian, U.S., E.C., etc.)
- Guidance documents (Guides, Guidelines)
- Published articles
- Best of industry practices

- Policies
- FDA 483s of other inspections
- Current best internal practices
- SOPs
- Current issues
- Management expectations, issues

One other type of tool used by some GMP auditors is questionnaires (Figure 8.3). These differ from checklists in that the questionnaires are open-ended and do not have "built in" requirements. Questionnaires are useful when doing a general survey or when benchmarking firms.

Preparing a checklist involves knowing what will be audited and the audit approach that will be used. To develop a checklist follow the steps listed below:

Figure 8.3. Example of a questionnaire with open-ended questions.

GMP Survey Questionnaire
Medical Device cGMPs—Subpart B

Area surveyed: _____ Date: _____

Question [QSR-GMP Reference]   *Response*

1. Does the firm have a quality policy?
   [§820.20 (a)]
2. How is the quality policy communicated?
   [§820.20 (a)]
3. To whom is the quality policy communicated? [§820.20 (a)]
4. Is there a current organizational chart
   available? [§820.20 (b)]
5. How does the quality unit relate organizationally and functionally to other parts
   of the firm? [§820.20 (b)(1)]

---

1. Identify the type of audit and specific audit approach(es); use suggestions from the particular audit approach(es) as a guide.
2. From the applicable regulations and guidelines, list things to consider along with specific requirements such as SOPs, recordkeeping requirements, etc.
3. From recent publications and talking with subject matter experts, identify additional requirements.
4. Review policies, SOPs, and other internal documents for requirements.

Checklists can be developed in a word processing package or database and then selected, customized, and printed out for a particular audit. It is helpful to include a reference that is the basis for the requirement.

As you prepare the checklists, give some thought to the level of detail that is included, which primarily depends on the experience and knowledge of the auditors. Newer auditors would need more detail; expert auditors would need less. In deciding how much detail to include, think about who your auditors may be in the next year or two; for training and consistency reasons, you may want to increase the amount of detail. Checklist items can be presented as either questions or statements, depending on the preference of the organization.

In the comments section, the auditor might write in specific examples of situations or documents that do not meet the requirement. Also, situations that display excellence can be elaborated upon. The general formats are shown in Figures 8.4A–D.

Documents to Request from the Auditee

To save time, you should request certain documents from the auditee that you will review *before* the visit. These can

Figure 8.4A. General format for a checklist.

Item No.	Requirements with desired level of performance.	References
	Standard	
	Comments/notes	

Figure 8.4B. An example from a highly detailed checklist for deviations.

1.	Deviations are investigated within 24 hours of their being observed.	SOP X123/3 (3.1)
	OS ❑ A ❑ MN ❑ MJ ❑ CR ❑ NA ❑	
	Comments	
2.	The deviation investigation is conducted in a timely way and is completed within 30 business days.	SOP X123/3 (3.3)
	OS ❑ A ❑ MN ❑ MJ ❑ CR ❑ NA ❑	
	Comments	
3.	The deviation report is clearly written following the standard format.	SOP X123/3 (3.4)
	OS ❑ A ❑ MN ❑ MJ ❑ CR ❑ NA ❑	
	Comments	

Figure 8.4C. An example from a less detailed checklist for deviations.

1.	Deviations are begun, investigated, and completed in a timely way, with reports written using the standard format. OS ☐ A ☐ MN ☐ MJ ☐ CR ☐ NA ☐ Comments	SOP X123/3 (3)

Figure 8.4D. Part of an audit checklist for examining training.

1.	Training records for each individual in the department are readily available and complete. OS ☐ A ☐ MN ☐ MJ ☐ CR ☐ NA ☐ Comments	US §211.25 (a); CA C.02.002/ 4.4 SOP 1748/2
2.	Training requirements for the past 4 quarters have been met for individuals in the area. OS ☐ A ☐ MN ☐ MJ ☐ CR ☐ NA ☐ Comments	US §211.25 (a); CA C.02.002/ 4.4 SOP 1748/2

help you understand the facility, organization, operations, and quality systems in place. If a contract acceptor (vendor or contractor) is not willing to furnish this information in advance, ask them to have it ready for your team to review on your first day. For internal audits, the audit team will have most documents already available.

The documents you request depend on the type of audit, its scope, and the audit approach you will be using. Some examples of documents to request are as follows:

- Organization charts
- Building layouts
- Policies
- Water system drawings
- SOP index
- Specific SOPs
- Validation protocols
- Validation reports
- Recent quality reviews

Note: An "unofficial" FDA guideline, *Foreign Inspection Guide*, was released in 1992 (FDA 1992). It provides a list of documents that FDA inspectors are told to request prior to an overseas audit. This may be useful to GMP auditors as well.

Reviewing Historical Files and Documents

It is very helpful for auditors to understand what previous auditors looked at and found, as well as what they did *not* look at. Also, the current audit team may need to follow up on commitments the auditee made to previous audit reports. Sometimes, historical files can be a gold mine of detail and save the current audit team time.

Action Plan

The action plan is your team's list of what will be done during the audit (Figure 8.5). It includes the responsibilities for each auditor, priorities, and a schedule. The plan, written for use by the team, will be followed during Phase II, when the team conducts the audit.

Figure 8.5. Example of part of an action plan.

Audit Action Plan

Location, date
Manufacturing and Packaging, Corp. Tech. Center, Oct. 8, 1997

Agenda, estimated time, responsibility
1. Kickoff Meeting (8:30–9:00): Frank, Janice
2. Manufacturing training activities (9:00–15:00): Frank, Viveca
3. Warehouse/distrib. (9:00–15:00): Janice, Julie, Ida
4. Auditors gather and review findings (15:00–16:00)
5. Oral comments with area managers and what happens next (16:00–16:30): Viveca, Ida, Frank

Prioritized Items
1. Individual training plans are defined.
2. Individuals are up-to-date in recording training.
3. Recent GMP courses (Basics, Documentation Skills) have been entered.

Future Plans
1. Planned date for reviewing findings from all audits in series, planning preparation of report: Oct. 17
2. Draft report written by Oct. 25–Viveca and Ida, primary writers.
3. Circulate draft: Oct. 25–31.
4. Finalize/distribute: Nov. 4.

Working Papers

Working papers are all of the documents required for an effective and orderly execution of the audit plan (and action plan). They include the audit and action plans themselves, checklists, auditor notes, and collected documents. All working papers will need to be identified with the name and date of the audit as well as who prepared them. At the end of the audit, they will be retained.

Communicating With the Auditee

Communicating with the group that will be audited is one other important task during the preparation phase. There will probably be several times when you talk with the key contact there, such as when you establish and confirm the audit dates and present the audit plan.

A frequent question is, "How much detail should be sent to the auditee in advance of the audit?" One position is that very little should be sent, as the auditors should see "real life" areas that have not been prepared in advance. The position on the other end of the spectrum is to provide the plan and checklist several weeks in advance of the audit team's arrival to help establish trust and better relationships.

Most auditees will not be able to make substantial, systemic changes after receiving the audit plan and checklist a month in advance. Even if they do, the auditors will see evidence of this, for example, logbooks with entries that don't go back very far, SOPs that have been recently written and approved, a flurry of training activity over the past weeks, etc. Auditors will also have access to how the area did things in the past by way of documents, batch records, and the like that give a historical perspective.

Confirming Phone Call

The week before the audit takes place (or before you leave your facility), it is prudent to phone the auditee just to make sure things are going according to plan. Auditees have been known to "pull the plug" on an audit at the last minute, causing embarrassment, wasted time, and travel money to audit teams.

Milestones for Phase I: Preparation

You are ready to conduct the audit if you know:

- What you are going to do,
- Who will be involved,
- The standards that will be used,
- Specific things that you will be looking for,
- What you will do with the information once you get it,
- Something about what you will be seeing, and
- That the team leader has communicated with the area to be audited, giving them information and a comfort level about what will be happening.

CHAPTER 9

Phase II: Conduct the Audit

Goals

- Distinguish between desk and on-site audits.
- Describe what happens during the on-site audit.
- Present a process for evaluating activities, system components, etc. that auditors can use.

Conducting the audit involves two primary tasks. First, you will do a *desk audit* to review the documentation that has been provided. Second, you will visit the area and conduct the on-site portion of the audit. In conducting the audit, you will be following the audit plan and the action plan you prepared in Phase I. You will also be using a structured evaluation process as you compare the situation against the requirements detailed in the checklist.

Audits or regulatory inspections are often considered *snapshots in time.* That is, the results of the audit depend on what the particular audit team saw at the particular time and day they were there. Auditors, however, can take a historical view of activities and decisions by way of the documents and records that the auditee should have readily available. This historical information can be used in determining if an observed deficiency is an isolated, one-time occurrence or if it indicates a systemic problem.

Who Is Involved

The team leader and audit team members are primarily involved in conducting the audit; the auditee along with its management will also be involved. The auditee will have one or more audit coordinators who will facilitate the auditor's visit.

When It Is Done

The desk audit is done one or two weeks before the visit whenever possible or, when required, on the first day of the site visit. The site audit is conducted on the dates that have been previously agreed upon.

The Evaluation Process

Auditors have a logical sequence of steps they follow in evaluating a situation, process, system, or facility. Stripped to its essentials, the process asks six questions:

- Is the system, component, activity, process, etc. (i.e., whatever you are examining) formally described, such as in an SOP, method, policy, or drawing?
- Does the description include all the appropriate requirements and specifications?
- Is it implemented per the description?

- Is it functioning adequately?
- . Have there been failures? Are they isolated or systemic? Have they been adequately documented and investigated?
- If there are deficiencies, what is the potential impact on the user of the product? To the organization?

The Desk Audit

The desk audit is conducted prior to beginning the on-site audit. The desk audit will provide the auditors with the following:

- An overview of the auditee's organizational structure;
- A look at how the auditee has defined and organized its GMP system and the related components;
- Descriptions of how the firm *says* it develops, produces, and controls its products and information; and
- An opportunity to evaluate what the firm says it docs with GMP, regulatory, and/or corporate requirements.

Occasionally, such as if an audit is made at a vendor's site, the desk audit occurs right after the opening meeting. Some auditors consider the desk audit as part of the preparation phase; it doesn't matter when it happens, as long as it is done.

The desk audit is sometimes called the *system suitability* audit, where the programs, systems, and activities described on paper (e.g., the SOPs, methods, policies, protocols) are compared to the reference standard (e.g., GMPs, regulatory submissions, contracts). The desk audit should be done in a conference room where all the materials can be spread out and used by the audit team members. Each team member should be assigned documents to examine. The audit checklists can be used as a tool in reviewing the materials.

The Site Audit

During the site audit (sometimes called an audit of *system conformity*) the team will:

- Determine if the auditee is actually doing what they say they are (i.e., comparing what is stated in their policies, SOPs, protocols, etc. to real life);
- Evaluate how well they are doing it;
- Determine if the elements are integrated into a system or if pieces are unconnected; and
- Make organoleptic perceptions (observe how things look, feel, and smell).

Site Audit Agenda

A typical agenda for the on-site audit is as follows:

- Opening or kick-off meeting.
- Collect the data.
- Understand the significance and cause of any deficiencies.
- Verify your perceptions.
- Share information.
- Wrap-up meeting.

The Opening Meeting

The opening meeting, sometimes called the *audit kick-off*, is usually the first thing that will happen. For external audits, there will be an introduction of the auditors and the auditee's management. The tone that is set at this meeting is important: the auditors need to present themselves as technically credible, fair, and professional.

Other points covered in this meeting are

- Goals of the audit,
- Standards that will be used,
- How the results will be shared and used,
- The schedule for the audit, and
- Questions and answers.

For the audit to be successful, the auditors and auditees must all agree on what will be covered, the criteria, and the standards. Ideally, any disagreements should have been discussed earlier, but sometimes issues arise at this first meeting. The lead auditor is responsible for resolving any problems. Some examples of issues might be a manufacturing operation that is shut down because of mechanical problems or a contract packager that is working on the product of another firm and, thus, you are not able to observe. Sometimes these are show-stoppers, but usually there are creative ways of working around the problem, like rearranging the audit schedule or spending more time examining documentation. However, if you are not able to observe something that is critical to accomplishing the goal of your audit, you may need to call your management to find a satisfactory alternative. Making a confirming phone call the week before, as described in the preceding chapter, can minimize these last minute surprises.

Occasionally, you will find an auditee that is not cooperative with the auditors. For example, one company prevented auditors from entering certain areas, hoping that the serious deficiencies in the areas would not be observed and reported. Unfortunately for the firm, the audit team had more time to scrutinize documentation and discovered critical problems. This resulted in a scorching audit report.

At the opening meeting, observe who is present from the auditee's management. If senior management representatives are there, it indicates there is a high level of interest in the audit.

Collect the Data

Data are collected according to the action plan, using the checklists as a guide. If the review of documents has not been done during an earlier desk audit, this would be the first thing to accomplish, as it will quickly give an overview of the operation and their quality system.

In most audits, particularly if you have not recently seen the operations, it is helpful to understand the "big picture" of the process. Start at the beginning of the operation and look for the following:

- Inputs—materials and information,
- Transformation (the activity itself),
- Controls in place (including documentation), and
- Outputs—materials and information.

Most auditors find it easiest to follow the flow of raw materials and components from the receiving warehouse, then into dispensing, manufacturing, packaging, and finally into the finished products warehouse. Be alert for maintenance and in-process testing areas as well as utilities (e.g., air handling systems, deionized and purified water systems). You are looking for situations that do not meet the checklist requirements or general GMP expectations. In general, see how the firm has established ways to maintain and promote product

- Safety,
- Identity and everything significant used to produce it,
- Strength (or fitness for use) through its shelf-life,
- Purity, and
- Quality.

Experienced auditors develop some tricks that they use as they walk around. These include the following:

- If a surface is dusty, make a mark in the dust with your finger and come back towards the end of the visit to see if it has been cleaned. If not, it indicates poor housekeeping and the potential for cross-contamination.
- Look in wastepaper baskets and see what is being thrown away. You should *not* see any labels, controlled documents, or raw data.
- If you use the restroom, observe how many people do or don't wash their hands while you are in there. This could be related to training or reinforcements on personal hygiene.
- Continually look at logs and chart recorders to be sure they are properly identified and completed using good documentation techniques. Poor documentation can relate to poor design of documentation into the job, lack of training and supervision.
- Examine the lightbulbs and cleaning trays in Insectocuters® (insect control devices). Look for how many insects are in the catch trays and how long they may have been there.
- If there is something on the floor or hallway, monitor how long it remains there.

These items are not necessarily critical deficiencies, but they do indicate the general GMP awareness people in the facility have.

Talking with People

Much of the data that you collect will come from conversations with people. Some auditors find that interviewing personnel is time consuming and that they can get much of the same information from reviewing documents.

Interviewing does take time, but results can provide a different picture from what is simply written (or not written) down. Line personnel might tell you how things are "really" done, sometimes using a method that is very different from the approved procedure.

Chapter 10 provides more details about interviewing people and approaches that can be useful.

Looking for Loose Threads

An FDA investigator informally explained how he approached GMP inspections. In essence, he said:

> I consider a company and its quality systems to be a piece of fabric. In my inspection, I want to find out how good the fabric really is—how much structural integrity it has. During the inspection, I try to find loose threads. When I find one, I start pulling. If the quality system and its components are well constructed and properly integrated, pulling one thread won't have much of an effect. If the quality system components are not properly developed, in one or two pulls, the piece of fabric will disintegrate in front of my eyes.

As a GMP auditor, you also are looking for loose ends. Where are you likely to find them? First, use the general and specific items listed in the various audit approaches (Chapter 6) that you have included in your audit checklists. Second, you might want to use a *trace-back* technique to see if all the related pieces are there and tell the complete story. Using the SOP as a reference, you will also be determining if the activity was done as defined. Some examples of different trace-back methods include the following:

- Identify a lot number from a product made in the past year, and ask to see its history record (i.e., batch history record or device history record). Find the connections back to the raw materials and components used, testing/acceptance records, any deviations that occurred, and identities of personnel involved in producing, testing, and approving the product.
- Identify all personnel in a department, and examine their training requirements, training records, and qualifications to perform their jobs.
- Find a deviation or complaint, and examine the investigation that was conducted on it (or the rationale of why an investigation on the complaint was not conducted). For deviations that occurred during the manufacturing process, examine the dates to determine if the deviation was resolved before the lot was released to the market. See if the problem or complaint recurs in other lots or products.
- Find a complex table or graph contained in a submission to a regulatory agency, and examine all the raw data used in it.
- In the laboratory, find a recent result, and reconstruct how it was obtained. Look for calibration of the instruments and preparation of the reagents; see if there were any out-of-specification (OOS) results, and note the training of the laboratory personnel involved.

Sampling Issues: How Much? How Many?

An important question that many GMP auditors have relates to sampling, that is, how large a sample needs to be taken to show there is or isn't a problem. Some quality auditors will use a "decision sampling" statistical approach (Mills 1989). Other auditors will let their experience and instincts guide them.

Part of the sampling size decision depends on how much evidence you will need to support your decisions. For example, if you are collecting evidence to be used in a legal proceeding (as a regulatory investigator might do), you will want to collect all the supporting pieces you can. The more you have, the tighter the case. On the other hand, if the audit sponsor and the auditee do not need a great deal of convincing that there are problems or if the auditor has a large amount of credibility in the eyes of the sponsor and auditee, lesser amounts of data would support the audit results. In this situation, you will want to sample widely enough to determine if the problem is isolated or systemic in nature. As you collect the data, remember what a statistician told a group of students: one occurrence is simply an event; two occurrences is a coincidence; three is a trend; and four occurrences mean you have a real mess on your hands.

As you are planning the audit and during the opening meeting, try to get a sense of how much data the audit sponsor and auditee will need to be motivated to take corrective action.

Recording Your Findings

Clearly documenting your findings and collecting example documents will help your team analyze the data and prepare the report. It also supports your observations.

Copies of documents you collect should be clearly identified and dated. *If copies of documents are not available,* such as if a contract acceptor will not give you a copy to take with you, you should reference the name, ID number, date, and/or revision level of the document in your notes.

You may want to write your findings directly on the checklists in the "comments" section. Alternatively, some auditors write their findings in bound, prenumbered laboratory-style notebooks; one notebook may be used for many audits.

One caution as you document your thoughts: take your time. One of the worst things that can happen as you later try to decode your notes is to not understand them because they weren't written clearly or legibly. Some auditors will spend an hour or so each evening summarizing their findings and write more extensive observations in a lab notebook or on a laptop computer.

Many firms in the industry have been audited by an FDA investigator who arrives with a laptop computer to use during the audit. Company officials may (jokingly) ask if the PC is validated. The investigator will typically say it isn't, because it is used only for analysis and any decisions will be based on written data the investigator keeps in his or her journal. In other words, the computer is a secondary tool that supports the investigator.

Discussing Findings With the Team Members

At the end of each day, it is useful to share your findings and thoughts with the other members of the audit team. They may have information that will support whether the deficiencies you observed are isolated or systemic.

Understand the Significance and Cause of any Deficiencies

If you observe a situation that does not meet the expectations, you next need to classify it in two ways. First, what is the *scope* of the observation—is it an isolated occurrence or systemic failure? What is the potential magnitude of the problem? For this, you may need to repeat the technique you used to discover the original observation and observe if a similar problem is seen. As in other types of problems, it is relatively easy to determine if the deficiency is systemic—you should see a number of related problems. If the problem is only an isolated occurence, few related problems will appear.

Second, what is the *impact* of the observation—could the fitness for use, or the safety, identity, strength, purity, or quality (SISPQ) of the final product be compromised in any way? Are people at risk?

Identifying the cause of the deficiency may or may not be part of your goal as an audit team. For example, if you are evaluating a potential new supplier of components, the goal of the audit might be to determine if the supplier is suitable or not. If the supplier has problems, your organization doesn't really want to know why; they want to know if they should use the supplier or not. On the other hand, if you are conducting an internal GMP audit or auditing an affiliate of your company, one of the goals would be to "make recommendations for improvement." To do this, you will need to understand the root cause of the problem. Some analytical techniques are discussed in Chapter 11.

Your answers to these questions will determine whether the deficiency is critical, major, or minor. Document your findings using the checklist and in your audit notes, include product identity, lot number, department, and line. Be as specific as you can be.

Verify Your Perceptions

Before you announce your observation, you must be sure your perception is accurate. For minor or major deficiencies, confirm your observation using two sources, such as documents or people. For major or critical deficiencies, try to "triangulate" with at least three sources, such as several different documents and/or people.

Your auditing and technical credibility will be enhanced if you can get others to notice the problem and its significance. Asking questions about the situation and its ramifications is useful here. *If you observe a problem happening*, such as a blue tablet (a "stranger") during the filling/packaging of a batch of white tablets, immediately point out the situation to your contact and those in the room, as they will be able to corroborate your observation. Watch for their reaction and what happens next. Are they taking the appropriate actions? Immediately record your observations.

As you collect information, record the information in your audit notes, being as specific as possible.

Share Information

One of the primary rules of auditing is to not surprise the auditee. As you verify your observations and their significance, share your findings with the primary contact and key people. This builds trust and gives them the opportunity to respond and correct the problem.

If the auditee makes corrections immediately, document this in your notes. (Observe, however, if they have addressed the problem on an isolated or on a systemic level; is the correction only a bandage or one that goes to the root cause of the problem?)

Some auditees request a daily informal meeting when the audit team presents their observations for the day. If management is present and involved in these meetings, it again is a good indication they are concerned about GMP and quality issues.

The Wrap-up or Closing Meeting

The wrap-up meeting is held at the end of the audit, when the auditor or audit team presents the preliminary findings to the auditee. If there is open communication with the auditee's contact during the visit, there should be no surprises at the final meeting. The auditee might not like the results, but at least the results are not unexpected.

Audits of current or proposed suppliers/contractors are slightly different from internal audits, as suppliers/contractors will want to immediately know if they will be (or will continue to be) an approved supplier. In some organizations, the audit team has the authority to do this. In other firms, the most the lead auditor can say is what their recommendations will be in the final report.

Note: Before the meeting, the audit team members need to do at least a preliminary analysis of their observations as described in Chapter 11.

The usual format of the wrap-up meetings is for the auditors to do the following:

- Restate the purpose and scope of the audit;
- Thank people for their time and assistance;
- List the area(s) that were examined;
- State the limitations (e.g., time, what was/was not available to be seen);
- Present areas of excellence;
- Present the initial findings (e.g., deficiencies);
- If there are deficiencies, tell what needs to happen to make them acceptable;
- Sense the reaction of the auditees;
- Be sure observations and details are correct before "exposing" them in the written report;

- Describe what you learned from them; and
- . Discuss what happens next with the audit report and follow-up.

During this final meeting, take notes of any commitments the auditee makes or issues that arise. Some of the observations you make during your audit may have been corrected by the end of your visit. The question then is, are the deficiencies mentioned in the final report?

Different auditors and organizations give various answers for this. Some groups say that the fact the problem existed at all before being called to the auditee's attention means that there is a deficiency. Another view is that sometimes "stuff happens" and that isolated situations that are corrected immediately should not be included in the final report. For example, if the audit team should notice one burned-out ultraviolet lamp in an Insectocutor® that is immediately corrected, it shouldn't be held against the firm. On the other hand, if the auditors find bulbs burned-out in many devices, it indicates that there is a systemic issue related to maintenance and pest control that needs to be raised.

If an issue is corrected during your audit, be sure the correction was correctly done (i.e., it was documented as appropriate and the change went through adequate levels of change control). One other aspect of this is if the auditee addressed only the specific situation that was pointed out to them or if they thought more broadly about the problem and made a systemic evaluation and correction. Some auditors will mention the deficiency in their report but, simultaneously, compliment the auditee on quickly correcting the situation.

Milestones for Phase II: Conduct the Audit

You are ready to analyze the audit data if you:

- Accomplished the audit and action plans;
- Completed at least the high priority items;
- Have an understanding of how things really are;
- Know if quality system elements exist, if they are implemented, and how well they are working;
- Understand how the auditee's operations compare against the standards;
- Have documentation to support your positions; and,
- Have team credibility in the eyes of the auditees.

Interviewing Auditees

Goals

- Describe an *interview*.
- Define *success* from the points of view of the individual and auditor.
- Identify and discuss factors that can influence an auditor's perception during an interview.
- Identify and describe three components of an oral communication.
- Discuss how to build better rapport with an interviewee.

As described in Chapter 9, when you conduct the audit, you will collect information by reviewing documents, observing operations, and talking with people.

Many people are nervous when talking to an auditor, because they realize the impact that their answers may have on the final outcome. Sometimes, their nervousness gets in the way of communicating. As the auditor, you will need to use some special skills to obtain the information you want.

What Is an Interview?

Interviews are purposeful conversations between two or more people. *Purposeful* means the conversation has a goal—it is trying to accomplish something specific. Interviews are *not* social conversations, in which the topics wander from the weather to family matters to vacations to sporting events. An interview also is not an interrogation, in which the interviewee is subjected to excessive pressure, hostility, or harassment.

For an interview to be successful, the auditor (i.e., interviewer) should have a goal in mind. For example, an auditor might want to:

- Understand the rationale for the sampling schedule of a high-purity water system,
- Learn additional details of how a lab method is performed, or
- Hear from different personnel their own understanding of what constitutes a deviation.

As you see, each of the examples is specific; questions that are focused help to accomplish the goals of the audit. A successful interview, for the auditor, obtains answers that will accomplish the audit goal, which is much larger in scope. The auditor might find out more than he or she initially expected, e.g., that the schedule for taking water samples recently was changed, but it is not yet included in the current SOP.

Interview Success

Success in an interview depends on the particular point of view that is taken. At times, success for one group or person might be a failure for another.

From the point of view of the interviewee, *success* means appearing competent, professional, and correct. At a very basic level, it means not appearing foolish in the eyes of the interviewer.

For the auditor, *success* in an interview means having his or her questions answered. The questions relate to the various points on the audit checklist; the answers are used to accomplish the goal of the audit.

The auditor is not looking for the "right" answer during an interview (or when reviewing documents). He or she is trying to determine the facts.

The First Critical Seconds of the Interview

If you read almost any book on interviewing or public speaking, you will find it confirms what all of us experience: people make judgments—positive or negative—about others in the first minute or two of contact. (Some authors say this happens within seconds.) As an auditor, immediately establishing a positive rapport between yourself and the auditee is one of the most productive things you can do.

Some auditors take the position that it doesn't matter what the interviewee thinks of him or her—they are not there to be liked, just to get the facts. Another view is that if a good, "safe," mutually respectful rapport is established, the interviewee will be much more comfortable in answering the questions.

The first moments of the interview can also color the auditor's perception of the interviewee. For better or worse, the auditor will immediately make some judgments about the interviewee's attitude, professionalism, and self-confidence. While frequently these initial judgments are correct, they are sometimes wrong. Auditors need to be careful to not jump to conclusions.

Auditors, like all of us, are also subject to bias as they visit facilities or talk with people. For instance, an auditor visiting a contract manufacturer might have a bias against generic-drug manufacturers and a bias towards name-brand firms doing contract work. To overcome this, it is critical for the auditor to first acknowledge to him- or herself that the bias exists, and secondly, to work so that the bias does not influence any decisions.

Spoken Communication

According to Mehrabian (1981), spoken communication, such as that occurring during an audit interview, has three components:

- Verbal—the content or information presented;
- Vocal—the way the information is spoken, e.g., sounds, inflection, volume; and
- Visual—what people see during a conversation, e.g., facial expressions, gestures, posture.

The more congruent these factors are, the more believable the speaker is. That is to say, the components of spoken communication can reinforce each other if they are working together. On the other hand, if they are given as "mixed signals," the message will be perceived as inconsistent.

If an inconsistent message is sent out, what do people believe? Again, according to Mehrabian's research, believability is based on:

- 7 percent of the verbal component,
- 38 percent of the vocal component, and
- 55 percent of the visual component.

To put this another way, if the interviewee's speech and actions aren't consistent with what is being said, much more weight will be given to the vocal and visual components.

This has implications for both the interviewee and the auditor. Interviewees need to be sure that the components of their communication—verbal, vocal, visual—are consistent and reinforce each other. For the auditor, if you sense an inconsistency, it *could* mean something.

Other Visual Cues

Some auditors would like to go further and attach meaning to everything an interviewee does (or wears) during an interview. If the interviewee has a nervous tick, does that mean that he or she is hiding something? Or, if he or she mumbles, does that mean they are not really confident in what they are saying?

There has been a great deal written about "visual" language and the hidden, unconscious meanings behind behaviors, but in the words of Freud, sometimes a cigar is just a cigar.

Behaviors that *do* seem to have more support as reliable indicators are as follows:

- Eye contact is important in connecting with a person;
- Dilated (wider) pupils suggest a person is more interested;

- Good posture and movement relate to higher self-confidence and projecting that confidence to others;
- Gestures can visually reinforce a verbal point;
- Body position and movement create the impression of energy on the part of the speaker; and
- If one person is inappropriately physically close to another, it can be perceived as threatening or harassing.

As you talk to people during an audit, you may meet some who are nervous. Quickly establishing positive rapport with these interviewees can help them provide you better information. To do this, first put the person at ease. Explain who you are and what you need. (If you are auditing a vendor or other external organization, ask your host to introduce you, as it will indicate to the person that the firm endorses your presence.) Try to set the initial tone of the meeting as more of a conversation than a formal interview.

Some other ideas for working with reluctant interviewees include the following:

- Treat the interviewees as the experts—you are there to learn from them.
- Ask them to show you things and describe what they do.
- Ask open-ended questions that are specific, such as:
  - What do you do in your job?
  - How do you know . . . ?
  - How do you document . . . ?
  - Can you tell me about . . . ?
  - What if . . . ?
  - What would you do if . . . ?

You may encounter an interviewee who uses phrases such as "*We always* . . ." or "*every time* . . . ." These broad generalizations can indicate. a bluff on the part of the speaker. In this situation, ask the interviewee for *specific examples* of when this situation actually happened.

Experience has shown that the more comfortable and trusting people feel with you, the more they will share with you. Sometimes, however, you will meet someone who, for whatever reason, wants to give you very specific "insights" into the organization. Be wary in these situations, as the person may have a not-so-positive reason for sharing all of these things; weigh the comments very carefully before acting on them.

At other times during an interview with an auditee, you may get a "hunch" about what is being said or shown you. These visceral ("gut") reactions can be important. They can mean that you are sensing something on a level that you cannot (yet) consciously quantify. Just as you should not disregard these hunches, you should not impulsively act on them.

For example, you are presented a batch manufacturing record for a recently made product. You look through it, and it looks very good. There are very few corrections; the pages appear almost pristine in quality. The manager you are talking to is very convincing in his statements, and he overwhelms you with the self-confidence he exudes. As you walk away from the meeting, you think that everything seems great, but as you replay the conversation in your mind, you begin to get a strange feeling about it all.

In a case like this, first, don't trust your initial impression that everything is fine. Listen to your instinct. Second, look for congruity with what your experiences have been. You know that if a batch record is used in production, they *look* used; they are not perfectly clean. Third, check the facts out with other people or documents. Request an interview with the operator who used that batch record. Ask him or her how they are able to keep a working document so clean. Finally, don't let your wishes cloud your thinking. You are interested in discovering the facts, not in supporting a predetermined point of view.

Conclusion

Interviewing is an important aspect of a GMP audit. Interviews take time, but when properly done, they will give you information about an organization that records and documents cannot. Interviews are situations where *each side* observes and evaluates the professionalism, expertise, and attitudes of the other. For many auditors, interviews provide one of the experiences they most value in auditing as they meet new people and continually learn new things.

CHAPTER 11

Phase III: Analyzing Audit Data

Goals

- Describe why analysis is done.
- Describe the goal of analysis.
- Identify techniques to analyze data as you are collecting it during an audit.
- Identify techniques to determine if deficiencies are isolated or systemic.
- Discuss particular issues related to analyzing data.

Analysis of audit data means organizing it in such a way that you are able to obtain meaningful information. Experienced auditors do much of this during the course of the audit, as they use observations to develop and test hypotheses as to what are the underlying causes of the problems.

Who Is Involved

Analysis of the audit data is done by members of the audit team.

When It Is Done

As the audit team prepares the report, the auditors will analyze their findings. Some of this will occur prior to the presentation of the preliminary, oral report at the audit's close-out meeting, with a more thorough analysis as the written report is prepared.

In the audit model, we have established analysis as a separate phase in order to emphasize its importance and highlight how it can be done. Analysis does not have to be a long, formal process. It is simply looking through all of the audit observations and determining the real significance of the individual data points.

Frequently, you will find "themes" in the data where many observations relate to a general GMP concept. For example, an audit team may have found the following major deficiencies throughout the facility:

- Operators did not consistently follow the line clear-ance and cleaning procedures for packaging equipment.
- SOPs were not available to those needing them.
- SOPs had not been reviewed or updated within the past two years.
- SOP books were not maintained by anyone; there was no one assigned to this task.
- Training on tasks was done informally; SOPs were not used as the basis for training.

All of these observations relate to SOPs. Looking at them together, the audit team can identify the more general theme: *There is no systemic way of reviewing, updating, distributing, maintaining, and using procedures.*

Providing the auditee with this type of information is much more useful than just the individual observations, as it gives the auditee a big picture view of the issue, which is supported by the data.

The concept of audit themes is related to the Pareto Principle of the "significant few and the insignificant many." In other words, most of the deficiencies found in the audit are caused by a few root problems.

Analysis While Collecting Data

Figures 11.1–11.3 are examples of forms that can be used to collect and arrange data. With these tools, problems such as a specification that was used (or referred to in a manufacturing document) before it was approved, or an SOP that was referenced but doesn't exist, will become

Figure 11.1. Chart for tracking events, follow-up dates, and closures.

Summary of Periodic Product Quality Reviews
Conducted Jan. 1/94—Jul. 30/97

Product name	Most recent date completed	Previous PPQE completed	Essential items reviewed?	Recom. made?	Recom. followed up?	Closed?
ASA tabs	Oct. 3/96	none	yes	no	NA	yes
ASA enteric coated	Oct. 3/96	Jun. 2/94	yes	yes	no data available	no
Eryth	Jun. 30/97	Jun.15/96	yes	yes	yes	yes

Figure 11.2. Example of chart for tracking events.

Deviation Tracking (Deviations occurring between Jan. 1—Dec. 31/96).

Dev. number	Date observed	Date Dev. initiated	Date report approved	Lot No(s) involved	Date(s) lot(s) released
96001	Jan. 4/96	Jan. 4/96	Jan. 12/96	TA691B	Jan. 20/96
96002	Jan. 6/96	Jan. 6/96	none found	TA781C	Jan. 19/96
96003	. . .				

Figure 11.3. Example of chart to track expected/actual sampling and testing events.

Stability Sample Review
Lots expected to be found in program

Lot No.	Mfg. start date	No. expected	No. originally obtained	Reason*	Samples pulled at approp. time pts?	Samples run within 30 days of time pts?
T2362	Feb. 23/95	5 x 20s	5 x 20s	N	Yes	Yes
T2351	Feb. 27/95	5 x 20s	5 x 20s	C (C#95-7)	Yes	Yes

* Reason codes: (N)ormal monitoring; (P)roblem lot; (D)evelopment lot; (C)hanged lot (CC#)

obvious. Depending on what you are examining during the audit, you may want to create other data collection forms. Things to keep in mind as you do this include the following:

- Keep them simple so problems readily appear.
- . Develop the form so you can see if deficiencies are isolated or occur in patterns (which might give a clue to the root cause of the deficiency).

Isolated vs. Systemic Deficiencies

Understanding if a deficiency is isolated or systemic is important because it can indicate the scope of the problem. For example, an incorrectly written lab method used for release testing of a raw material could affect all products that include that material. On the other hand, if a lab analyst makes an incorrect final dilution (usually an isolated situation), the error would be much more limited. Table 11.1 gives characteristics distinguishing isolated from systemic deficiencies, and Table 11.2 provides additional examples of both kinds of deficiencies.

Figure 11.4 shows a tool that an audit team can use to gather their individual findings to see if deficiencies are common to other areas. Depending on the detail needed, individual requirements can be written (i.e., fine detail) or several requirements can be included together. High

Table 11.1. Distinguishing Characteristics of Isolated and Systemic Deficiencies

An **isolated** deficiency . . .	A **systemic** deficiency . . .
• Rarely happens • Tends to happen randomly throughout the organization • Will show no meaningful pattern	• Happens more than once • Could be connected to a particular product, process, task, material, time, person, or organizational unit • Will show a pattern

Table 11.2. Examples of Isolated and Systemic Deficiencies

Isolated	Systemic
• Operator or lab techni-cian spills a sample • Wrong expiration date written on a reagent bottle • Latex gloves rip • Motor on tablet com-pressing machine burns out	• SOP contains an error • Batch record is poorly written so instructions can be confusing • Area management does not reinforce requirement to continually record informa-tion properly • Preventative maintenance program does not include tablet compressing machine motors

levels of detail can be useful in pinpointing weaknesses in a system.

One method used in the creation of checklists and for analysis is that of a "trigger" question written at a low level of detail. If auditors find that this requirement isn't met, they then get more detailed and specific in follow-up questions to isolate the specific deficiencies. An example of a low-detail requirement is "SOPs are properly prepared, controlled, and distributed."

Examples of fine-detail requirements are listed below:

- SOPs are properly generated.
- SOPs are adequately reviewed and approved.
- SOPs are adequately controlled.
- SOPs are available to those using them.

In the U.S. pharmaceutical industry, one set of cat-egories used to classify failures or deficiencies is *process-*

Figure 11.4. Chart for tracking deficiencies between different areas.

Deficiency Element		Area/Product				
Item	Description	Mfg Solids	Mfg Liquids	Pkg Foils/Blis	Pkg Bottles	Main. Engin.
5.1	SOPs not approved	X	XX	XX		XX
5.2	Current version not available	XX				X
6.4	Training plans not available					XXX
6.6	Training records not up-to-date	X				XXX

X   Isolated deficiency (one or two incidences)
XX   Frequent incidences of deficiency
XXX   Element is totally deficient

related and *nonprocess-related* (Wolin 1993). Summarized, *process-related errors* are those where there is something inherently wrong with the defined process or SOPs. In other words, even if people followed the process as written, it will lead to problems.

A *nonprocess-related error*, in Judge Wolin's written opinion, would be an operator error. Here, the operator makes a mistake by not following a procedure correctly or by making an inappropriate decision.

Process-related errors are an example of systemic failures. Nonprocess-related errors can be isolated failures, or they could have systemic roots. For example, all of the failures are related to one person not being capable of correctly performing the tasks or management is encouraging people to take risky shortcuts. Table 11.2 identifies examples of isolated and systemic deficiencies.

Other Analytical Tools

Your audit team may be asked to provide recommendations or help solve problems as a part of its audit goal. There are some useful models and tools that may help.

Performance Model. Figure 11.5 shows some of the typical barriers to proper performance that can apply to both isolated and systemic failures. This model can be used if people are identified as the primary cause of a failure. Training, which is frequently identified as something to prevent problems, should be considered only in conjunction with the other categories listed. Another useful performance model that is built around a flowchart for decision making is found in Mager (1984).

Cause-and-effect chart. Sometimes called a fishbone diagram, this tool is useful as a problem-solving tool to identify potential causes of a problem. It can also be used to show relationships between a problem and its root cause.

Figure 11.5. Performance model (Vesper 1993).

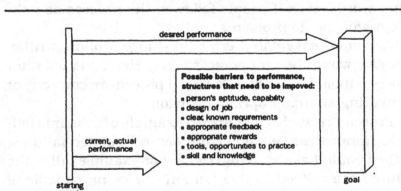

Additional graphical and statistical tools that might be useful to audit team members can be found in *The Memory Jogger*™ (Brassard and Ritter 1994).

Self-Analysis by Team Members

One last activity that is very useful for continually improving the GMP audit program at your firm and enhancing the skills of even the best auditors is to take a little time after the audit to reflect about how it went. This is not meant to be a soul-wrenching task, but rather a look back on how you would do it better next time. Some simple questions that might be useful include the following:

- What went very well/smoothly during the audit?
- What did not go well/smoothly?
- For each phase in the audit cycle, what would we do differently next time?
- What did we learn?
- What were we not prepared for?
- What value did we provide to the audit sponsor? To the auditee?

One other method for evaluating the performance of the audit team that some firms use is a questionnaire that is given to the auditee. Topics covered include professionalism, listening skills, attitudes, communication, and knowledge. Completed questionnaires are returned by the auditee directly to the head of the quality unit.

Milestones for Phase III: Analysis

You are ready to report your results when you have:

- Separated the "significant few" deficiencies from the "insignificant many,"

- Identified whether the deficiencies are isolated or systemic to the areas audited,
- Identified themes related to the deficiencies, and
- Identified ways for your team to perform better next time.

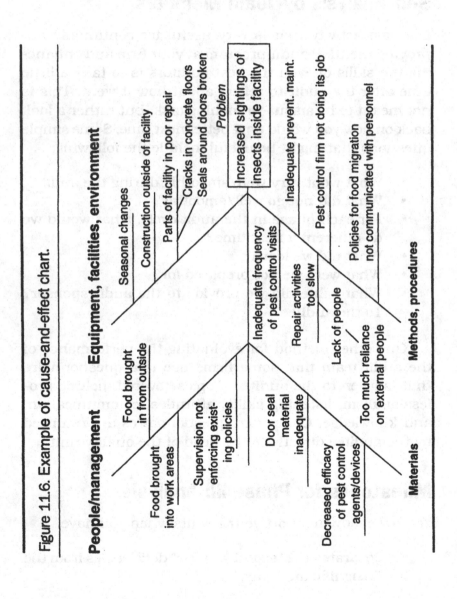

Figure 11.6. Example of cause-and-effect chart.

CHAPTER 12

Phase IV: The Audit Report

Goals

- Identify readers and purposes of an audit report.
- Identify what is typically in an audit report.
- Discuss variations in an audit report useful for certain audiences.
- Describe and provide examples of comparison reports.

The audit report phase includes making a preliminary, oral presentation to the auditees and preparing and distributing a detailed written report to management (or the audit sponsor).

Who Is Involved

The audit team, with management members to review the draft document, are the primary individuals involved in preparing the audit report. Sometimes, auditee management is asked to review the draft report as well.

When It Is Done

The oral presentation of the preliminary audit findings is done during the wrap-up meeting with the auditees. The detailed report is written at the end of the audit and provided to local management soon after. The audit team should have a goal to provide a draft report for review within a week of completing the audit and the final report within 2–4 weeks of the audit.

The report of a GMP audit, like other technical documents, serves many purposes and is read by different audiences at various times. For example, some of the purposes of an audit report include the following:

- Describing what was accomplished during the audit;
- Helping the auditee improve its operations;
- Providing information to the sponsoring management;
- Providing a formal assessment to the auditee;
- Being the basis for decision making;
- Providing a record of where the auditee is on its "quality journey";
- Demonstrating compliance with the audit procedure and policy;
- Helping the auditee justify changes, such as capital improvements and hiring additional personnel; and
- Providing a reference for future auditors.

Audiences, or those who might read an audit report, can include the following:

- Local and corporate management,
- Quality professionals,
- Purchasing agents,
- Technical operations management and staff,
- Product development scientists,

- Teams investigating deviations,
- . Consultants, and
- Training professionals.

As mentioned in Chapter 3, in some cases regulatory agencies have access to reports of vendors that a medical device manufacturer audits. With these many uses and audiences, reports need to be carefully written so they say what you intend for them to say.

Audit reports are written using a standard format to provide uniformity and comparability, as well as to speed the writing process. Analytical, organizational, and writing skills are critical in preparing a well-written document.

There should not be any "surprises" found in the report. In particular, there should not be anything of substance in the final report that was not presented orally at the close-out meeting with the auditees. One of the best ways to destroy credibility and a long-term relationship is to include items that were not discussed or to increase an observation's seriousness in the final report. If changes need to be made, the lead auditor should contact the auditee and discuss this in advance.

Standard Formats

Your firm may want to establish one or more formats for the reports and create style sheets in a word processing application that all audit teams can use. Table 12.1 describes some styles and audiences, and examples are provided in Figures 12.1–3B following the table. In addition, an example of a detailed formal report is shown in Figure 12.4.

The summary letter and executive summary are derived from the more extensive detailed report. The value of these is that they are targeted for specific needs, contain less material to read, and control the distribution of the detailed document.

Table 12.1. Various Types of Audit Report Used Within an Organization

Style	Audience	Description
Summary letter	Vendors, contract acceptors, internal auditees	A one- or two-page summary sent to contract acceptors that summarizes the goal, conclusions, results, and important findings. (Figure 12.1)
Executive summary	Internal senior management	A half- to one-page summary of the goal, conclusions, and findings of the audit that informs senior management of what took place. (Figure 12.2)
Detailed report	Internal (sponsor) management, quality personnel, internal auditee, etc.	A multi-page, detailed, formal report on the audit that provides the details of how the audit was conducted and all the findings. (Figure 12.3A–B)

Figure 12.1. Example of an audit summary letter to a vendor.

Chaos Drugs, Inc.
235 Lucky Drive
Batan Lake, TX 84525

January 23, 1998

Mr. Charles Lakewood
Director, Quality Services
Louie Containers
35234 Bissell Street
St. Louis, MO 45423

Dear Chuck:

On behalf of Chaos Drugs and the audit team that visited your
facility on January 11–12, 1998, we would like to thank you for the
time and assistance of you and your staff as we conducted an
annual GMP Vendor Audit of your facility.

As we discussed, we would like to summarize our findings in
this letter and reiterate our expectations so that Louie Containers
can remain a qualified supplier of folding cartons to Chaos Drugs.
Definitions of the audit classifications below can be found on the
attached sheet.

During our audit, which included an examation of your quality
systems, materials testing laboratory, warehousing, manufacturing,
printing/labeling, and shipping departments, we found the
following deficiencies:

Critical deficiencies: None were observed by the audit team
during our visit.

Major deficiencies:
- Proofreading of pre-press and post-press materials was not
  documented or dated by the proofreaders.
- Manufacturing records were not legible or complete;
  corrections made to errors were very difficult to decipher.

Continued on next page.

Continued from previous page.

- Manufacturing documents, including work instructions, specifications, and samples, were not consistently collected, stored, and maintained, making it difficult to reconstruct when an order was run.
- Samples of printed cartons are not defaced in a consistent way.

Minor deficiencies:
- Warehousing doors are left open for significant periods potentially allowing pests/insects to gain entry to the facility.
- Communication between the engineering and manufacturing departments provides only last-minute notification of an upcoming preventative maintenance activity, resulting in logjams and overtime work.
- Materials testing personnel are not consistently following your method LM254 dealing with destruction of sampled materials.

One particular area of excellence we noted was in the new laboratory equipment your firm has installed, along with its IQ/OQ. We were also impressed by the training your lab personnel have received in its operation.

As we discussed at the post-audit meeting, we will expect to receive a reply from you within thirty days from the date of this letter. In reviewing your response, we will look for how you will correct these problems and ways you will determine if the corrections have been effective.

Based on receipt of a satisfactory response, we will recommend to our purchasing department that Louie Containers be kept as an approved supplier.

If you have any questions, please feel free to call me.

Sincerely yours,

Chaos Drugs, Inc.
Lynn Shearson
Audit Team Leader
(702) 555-9000, ext. 231

Figure 12.2. Example of part of the executive summary sent to management on a monthly basis.

Summary of Vendor Audits
January 1998

Auditee	**Audit Leader**	**Audit Date**
Louie Containers	Lynn Shearson	January 11–12, 1998
35234 Bissell St.		
St. Louis, MO 45423		

Audit Type	**Follow-up Required**	**Next Audit Date**
Vendor audit—	Review of response to	First quarter, 1999
general GMP	letter; review of	
operations	changes at next normal	
	audit.	

The operations, management personnel/activities, and quality systems were consistent with previous audits with a noticeable improvement in general housekeeping. Deficiencies of previous audit were reviewed and had been corrected as per firm's audit response of Feb. 24, 1997.

Findings of deficiency
Critical: None observed.
Major: Included poor documentation of proofreading (it appeared to be done, but was not documented); poor documentation practices by manufacturing personnel; lack of systematic, consistent archiving of work records/samples (firm had difficulty in quickly accessing information requested); inconsistent defacing of printed matter.
Minor: Poor internal communication related to preventative maintenance; doors left open (possible entryway for pests); isolated cases of lab personnel not properly destroying materials.

Areas of excellence
IQ/OQ of new laboratory equipment along with training of lab personnel in its operation.

Conclusion
Pending response, there were no GMP/Quality reasons observed by the audit team not to recertify firm as a Certified Vendor.

Figure 12.3A. Example of an internal cover memo.

Chaos Drugs, Inc.
Interoffice Memorandum

Date: November 1, 1997
To: M.R. Goldman, Director, Technical Operations
From: F.W. Roussel
cc: I.A. Western, S.K. Spellman, S. Tracy

Topic: Audit Report, Technical Operations Training

Attached is the completed report of the audit conducted in
October 1997 of the training activities of the Technical
Operations Division.

We appreciate the assistance you and your colleagues have
provided to our audit team, allowing us to complete the audit
in less time than was orginally planned.

The issues described in the findings section are what were
presented at the wrap-up meeting on October 23. In particular,
we call your attention to the one critical deficiency listed. As
you are preparing your response (as per SOP QA-833), if
there are any questions or issues you would like clarified,
please call or e-mail me.

Figure 12.3B. Example of selections from an internal audit report.

This is a CONFIDENTIAL DOCUMENT. It is not to be released for any reason outside of Chaos Drugs without the written authorization of the QA Director.

Chaos Drugs, Inc.   Internal GMP Audit Report

Area audited: Technical Operations (Corris St. and Corporate Technical Center)

Audit dates: October 8–9, 1997

I. Executive Summary
A GMP audit, focused on training activities in the technical operations area, was conducted at the Corris Street and Corporate Technical Center locations. Critical deficiencies were found in both the engineering department (TC-345) and in the Corris Street incoming warehousing department (CD-144). Four major deficiencies were observed in several areas, along with several minor ones. Barriers identified that may be contributing to the major deficiencies were ambiguities in SOP 748/2 that were unclear about specific training data to be retained by individual departments. Recommendations to correct this particular problem include clarifying the requirements and having a divisional training coordinator (approx. 30% of a full-time position) to coordinate training documents.

II. Audit Plan
Goals
Determine compliance with SOP 748/2 [Training of Technical Operations Employees] and CP 15/1 [Corporate Policy—Training]. Identify barriers in complying with SOP and CP. Recommend solutions for improvement.

Scope
Technical operations training program.

Specific areas audited
Manufacturing and packaging at Corporate Technology Center; Engineering at Corporate Technology Center; Warehouse and Distribution Center, Corris St.; Quality Assurance at Corporate Technology Center; Materials Management at Corporate Technology Center.

Continued on next page.

Continued from previous page.

Approach
General GMP Operations, specifically Training.

Type of audit
Focused internal audit.

Audit team
Lead auditor: Frank Roussel
Team members: Viveca White, Janice Newmar, Ida Peel, Julie Brown

Auditee contacts
Area training coordinators.

Audit standards to be used
Five categories (defined in SOP 943/1):
- Outstanding
- Satisfactory/adequate
- Minor deficiency
- Major deficiency
- Critical deficiency

III. General Notes
All items described in the audit plan were accomplished. Safety and Environmental training requirements were also reviewed; however, specific job-specific requirements were not evaluated due to lack of expertise on behalf of the auditors.

IV. Audit Participants
Colleagues assisting in the audit as department audit contacts were the training facilitators in each area. These included the following:
- Manufacturing and packaging at Corporate Technology Center— Mary Stenum
- Engineering at Corporate Technology Center—Barry Steambottom
- Warehouse and Distribution Center, Corris St.—Rob Williams
- Quality Assurance at Corporate Technology Center—Marty Mackie
- Materials Management at Corporate Technology Center— Cortney Luvin

Continued on next page.

Continued from previous page.

V. Areas of Excellence
Of the areas audited, one showing particularly outstanding performance was the Packaging department (TC-711), where senior operational colleagues had prepared detailed orientation packages for temporary and new personnel. These include a "survival guide" with commonly asked questions, maps of the area, and a glossary of commonly used terms and abbreviations.

VI. Audit Findings
Because multiple departments were audited, the deficiencies are listed by department.
1. Manufacturing (TC-063)
Training records on individuals are not kept current on a timely basis.
Critical deficiencies: None observed.
Major deficiencies:
   1. There was frequently a delay of 3–6 weeks between a training event and the entry of the event into the corporate training tracking database.
   2. Records awaiting input into the database are not kept in a systematic way; they are found in multiple places, such as on top of file cabinets and in desk drawers of the training coordinator.
Minor deficiencies: None observed.

2. Engineering (TC-345)
Training plans have not been developed for engineering and maintenance personnel; training on GMP, job skills, and safety/environmental topics have not been provided to departmental personnel in the past 18 months.
Critical deficiencies:
   1. Training plans have not been developed for personnel.
   2. GMP training programs, such as corporate programs (GMP Basics, Documentation Skills and Techniques, GMPs and Your Job) have not been offered to department personnel.
   3. There is no evidence of training on Safety and Environmental topics.
   4. Training on SOPs used by personnel (e.g., SOP 325, SOP 104) has not occurred.
   5. There is no plan in place for training to be provided to new or temporary personnel.
Major deficiencies: None observed.
Minor deficiencies: None observed.

Continued on next page.

Continued from previous page.

VII. Conclusions

1. Most of the areas audited met most of the divisional and corporate training expectations. Engineering (TC-345) was the department that showed the most significant gap between expectations and practice.

2. Of the other deficiencies observed during the audit, several were related to SOP 748/2, specifically the requirement for retention of learner evaluations. All of the training coordinators interviewed had a different understanding as to what was to be kept (e.g., the original document, only the test scores, the fact that the learner had passed, etc.).

VIII. Recommendations

With regard to conclusion point 2, the audit team recommends that a task force of training coordinators, led by the Manager of Technical and GMP Training, clarify SOP 748.

IX. Expectations—What happens next?

In keeping with the Corporate Auditing SOP (SOP QA-833), you are asked to provide to the Lead Auditor a response to this audit report within 10 days of its issue date. To assist you, a standard template is available in the QA public file on the Ca-Comp Network. In your response, please indicate the following:

- Corrections made, a description of the correction, and the date completed;

- Corrections planned, a description of the planned correction, its anticipated completion date, and the name of the person responsible for the activity; and

- How you will know that the corrections made or planned will be effective.

The audit team will review your response and contact you thereafter to confirm the changes have been made.

If you have any questions or want to discuss various options, please contact the lead auditor, Frank Roussel.

Figure 12.4. Detailed formal audit report.

Chaos Drugs

101 Java Joe Highway ● Toronto, Ontario ● Canada ● Q2U 4T1
Phone: 1-800-ChaosRx ● Fax: 1-416-ChaosFx

Quality Assurance Vendor Audit Report
Convenient Cartons, Inc.
October 9–10, 1997

Executive Summary

An audit team from Chaos Drugs visited the manufacturing facilities of Convenient Cartons Oct. 9–10, 1997, to conduct an initial certification audit of the GMP and quality systems.

There were no critical deficiencies observed during the audit; however, two major deficiencies were noted. In addition, three minor deficiencies were noted.

At the end of the audit, team members met with Convenient Carton management, who said they will take immediate action to correct all major and minor deficiencies.

The audit team saw nothing from a GMP/quality perspective that would prevent Convenient from being listed as a Qualified Vendor.

Audit Plan

Audit goal: Evaluate CC from a GMP and quality perspective to determine if they should be a Qualified Vendor for Chaos Drugs.
Scope: Manufacturing, controls, testing, release, and quality practices related to cartons that would be produced for Chaos Drugs.
Type of audit to be performed: Contract acceptor (supplier) audit.
Audit approaches used: Supplier requirements, equipment/facilities, general GMPs, and deviations.
Areas audited: Manufacturing facilities, inspection, testing, quality program, warehousing, and shipping.
Audit team members: Piki Williams (lead auditor), Sam Turbot

Company Background Summary

Convenient Cartons has been in business for 19 years, with an increasing amount of business supplying pharmaceutical and medical device firms (60% of total business). The balance of their sales is primarily to the computer industry.

Continued on next page.

Continued from previous page.

Convenient Cartons has 189 employees, 130 in manufacturing related roles; 7 in quality roles; and the remainder in sales, management, and administration.

The firm is in the process of becoming certified to ISO-9002, in the next 6 months, using a RVP of Germany as their registrar. (See Attachment 2 for more detailed firm information.)

General Notes
All areas were observed in operation as the firm was producing cartons for a computer hardware manufacturer. Convenient Carton's management was very helpful and professional in facilitating the audit and answering all questions. All information requested was quickly provided.

Personnel from Convenient Cartons who assisted during the audit were:

- Lawton Kidder, QA Manager
- Shelly Summers, Customer Relations Manager
- Claire Castro, Manufacturing Manager
- Tanya Gottalotta, Warehousing Department Head

Areas of Excellence
The Chaos Drugs audit team observed two particular areas where Convenient Cartons demonstrated outstanding performance:

- Customer support, including investigation of complaints and order fulfillment quality metrics.
- Document control—an internal computer system provides electronically controlled documents and information to all personnel throughout the site. This includes manufacturing orders, statistical process control, and SOPs.

Audit Observations
The following observations were made during the audit of the facility on the dates listed above using an audit checklist (Attachment 3). With the exceptions listed below, other items examined during the audit were satisfactory. (Definitions for the standards below can be found in Attachment 1.)

Critical deficiencies: No critical deficiencies were observed during the audit.

Continued on next page.

Continued from previous page.

Major deficiencies: Two major deficiencies were observed during the audit. All are related to documentation and record-keeping practices.

- There is no documented follow-up to show that changes noted during the proofreading were corrected as requested.
- The sequential logs for lines 3, 4, and 6 were incomplete. Missing information prevented an observer from easily being able to reconstruct what products were made and when.

Minor deficiencies: Three minor deficiencies related to record-keeping practices were observed during the audit.

- Temperature and humidity recorders in stock supply area do not have fresh chart replacements weekly. Recorders observed show at least three weeks of retracings.
- There is no consistency in how samples are defaced. Some personnel use ballpoint pen, others pencil, and others wide-tipped markers.
- Films sent out for processing are not recorded in any type of log.

Conclusions

During the audit, the auditors saw that Convenient Cartons has a solid, functional quality and GMP program in place in manufacturing, control, warehousing, testing, etc. Documentation and record-keeping seem to be areas throughout the operation in need of attention and improvement.

The auditors saw nothing from a quality/GMP point of view that should prevent Convenient Cartons from being an approved supplier.

Expectations

As discussed at the audit wrap-up meeting, the QC manager from Convenient Cartons is requested to reply to the findings of this audit in writing within 10 business days of receiving it.

Chaos Drugs would expect to see how you plan to correct the situation, when the correction should be made, and how you know your correction will be effective.

Attachments [not shown]

Chaos Drugs standard definitions

Organizational questionnaire results, organizational chart

Sample of master checklist used for this audit

The Standard Audit Report

Your organization can delineate what specifically will be included in your standard report. Some typical headings are listed below.

- Name and location of auditee, date of audit.
- Executive summary—a one-paragraph summary of what was found and any conclusions made *vis-à-vis* the audit goal.
- The audit plan, including
    - audit goal, scope
    - areas or products audited
    - audit approach used
    - audit team members and leader
    - standards used
- General notes—what was not examined, changes in the plan, caveats in interpreting the report
- People involved from the auditee's firm or department (name and title)
- Areas of excellence—items noted during the audit
- Audit findings
- Non-conformance theme #1
    - Critical observations related to theme
    - Major observations related to theme
    - Minor observations related to theme
- Non-conformance theme #2
    - Critical observations related to theme
    - Major observations related to theme
    - Minor observations related to theme
- Additional non-conformance themes
- Conclusions (related to goal of the audit)
- Recommendations and expectations
- What happens next
- Attachments (as appropriate)

- Organization background, organization chart (for external audits of contract acceptors, contractors)
- Blank (i.e., unused) checklists with standards, references

If your organization is developing or revising standard report templates, create a report using dummy information or reformat a previously issued report. Have the document reviewed by all potential users, asking that they supply comments and suggestions. After they have approved the format, "lock it down" and formalize it in your auditing SOP. Include the template or style sheet as an attachment. You may also want to provide an electronic version of it directly to users or by placing it on a computer network.

Details on Particular Sections of the Audit Report

Most of the report is self-explanatory; however, added detail is provided for several sections.

General notes. Since the audit describes what happened, this section gives you the opportunity to discuss unusual events or changes from the audit plan or action plan. If circumstances forced a change to the audit plan, this should be noted here along with the reasons for the change. For example, an area was not available to the audit team or an audit was lengthened by two days because of serious findings. It is also very useful, so as to prevent future controversy, to note what was *not* audited (due, for instance, to time constraints).

Areas of excellence. It would be unusual, in any type of audit, to *not* find something the auditee was doing that was a novel, terrific way of meeting an organizational, quality, or compliance need. Pointing several of these out reinforces their value to the auditee and provides some balance to the report. It also can be a vehicle for sharing excellence with others in the broader organization.

Conclusions. This section of the audit report presents the bottom line as to what the observations all mean. The conclusion is crafted around the goal of the audit (e.g., whether the facility is prepared for a pre-approval inspection, if a contract acceptor meets the standards of being a qualified supplier).

Working Papers

As mentioned in an earlier chapter, working papers are the documents used and obtained during the audit, as well as the notes made by the auditors. In effect, the working papers become the evidence or raw data on which the observations and conclusions are based.

Keeping information organized and easily accessible is important as you collect and arrange all of the documents. Audit team members frequently are asked for details of a particular observation, and it is useful to be able to quickly connect an observation with the raw data. One way to do this is to notate a copy of the report with the page number, source document, or other identifier of the particular working paper.

As You Write the Report . . .

Some things to keep in mind as you write the report include the following:

- Keep the report as simple as it can be.
- Strive for clarity, objectivity, and directness.

- The audit report's quality (or lack of it) will reinforce (or undermine) the credibility of the actual audit.
- The audit report (and oral comments) not only have to present the deficiencies but also motivate those who will correct them.

Be very careful as you make conclusions related to regulatory or GMP compliance, as these must not only be factual but worded in a way to minimize the negative impact if pulled out of context. For those new to writing audit reports, have an early draft informally reviewed by colleague who is organizationally savvy. Some organizations require that drafts be reviewed by a lawyer before any official document is issued.

Proprietary and Confidential Nature of the Information

Internal GMP audit reports are generally off-limits to regulatory agencies like the U.S. FDA. Because they may contain sensitive, candid information that could be misused, it is important that the reports be considered confidential and/or proprietary information. The goal of this is not to restrict proper use of the information, but to limit distribution of the document only to those who have a genuine need for it.

Occasionally, firms have had their internal audit reports opened up in legal proceedings during discovery, such as during a product liability lawsuit. There are ways of protecting audit reports from this, such as having the audit be requested by the chief legal counsel of a corporation. If this may be an issue, discuss it with the appropriate legal experts.

Milestones for Phase IV: Report

You are ready to follow up and close out the audit when you have:

- Organized your findings by themes and supported them with critical, major, and minor observations;
- Identified notable areas of excellence;
- Made conclusions related to the goal of the audit; and
- Made references between the observations and the working papers.

CHAPTER 13

Phase V: Follow-up and Closure

Goals

- Describe what happens during the follow-up and close-out phase.
- Identify what should be included in the auditee's response to an audit report.
- Describe what a lead auditor should look for when reviewing an auditee's response.
- Given a particular situation, identify ways of following up on the audit responses.

Follow-up and closure is evaluating the auditee's response to confirm that the corrective actions (if required) are satisfactory and were accomplished. Management (i.e., the audit sponsor) is informed and all working papers are properly archived.

Who Is Involved

The auditee provides the response to the audit report with the lead auditor evaluating the response, filing all the related documents, following up, and preparing the closeout memo.

When It Is Done

Assembling and filing the audit-related documents is done immediately after the audit and continues until the closure memo is written. Following up is done according to the standard timeframe that was agreed to by the auditee and audit team.

Following up and closing out an internal GMP audit or a GMP audit of a contract acceptor is different from an ISO or other third-party quality audit. Stakeholders of internal GMP audits want to be sure that the organizations audited have corrected the deficiencies that were pointed out. Firms using vendors or contractors want to know that improvements have been made so that their requirements are being met. The follow-up and closure discussed in this chapter "closes the loop," completing one audit cycle.

In the case of ISO audits and audits where the auditors are consultants or other third parties, lead auditors are not involved with the follow-up; it is left to the audit sponsor to handle as it desires. The ISO auditing standard (ANSI/ISO/ASQC Q10011-1-1994) states:

6.0 AUDIT COMPLETION
The audit is completed upon submission of the audit report to the client [i.e., audit sponsor].

7.0 CORRECTIVE ACTION FOLLOW-UP
The auditee is responsible for determining and initiating corrective action needed to correct a nonconformity

or to correct the cause of a nonconformity. The auditor is only responsible for identifying the nonconformity.

What to Expect From the Auditee

The expectations the audit team has of the auditee should be defined in your firm's auditing SOP and communicated to the auditee during the wrap-up meeting and in the formal report. Typically, you will want the following:

- Confirmation from the auditee that they have received the audit report;
- Their response to the deficiencies, including how and when they will correct them; and
- If the corrections will take time to implement and the situation demands it, what temporary measures the auditee take.

It is important that the auditee reply to the report in a timely way, such as within two weeks. Figure 13.1 gives an example of part of a response to an internal GMP audit, as well as a useful format.

Figure 13.1. Example of a format that can be used when responding to an audit observation.

Item	Audit Observation	Our Response
1.	The current version of SOPs is not consistently available in the packaging department.	The department had not designated someone to maintain the SOP binders as other departments had. A clerk has now been assigned to this task; the task has been added to the job description.

Evaluating Auditee Responses

It is usually the responsibility of the lead auditor to receive and evaluate the responses. In addition to the points made above about what to expect from the auditee, there are some additional criteria you should apply when reviewing the responses. These include the following:

- Does the response show an understanding of the real issue and that the firm is addressing it?
- Does the response go beyond a "quick and dirty" correction?
- Does the auditee attempt to institutionalize the change, that is, apply the improvement to other lines, products, sites? (For vendors and contractors, this may not be practical because of the requirements of their other customers. See Chapter 14.)
- Does their solution go beyond simply training people? (See Figure 11.5.)
- Do the changes and controls make sense or might they cause even more problems later on?
- Are there any regulatory issues, such as agency pre-approvals or notifications that need to occur?
- If the issue has not already been resolved, do they give a reasonable time frame for doing so?
- If there were critical deficiencies that could affect products on the market, are they taking appropriate steps?

When Differences Arise . . .

While some audit responses will be satisfactory, you will undoubtedly review some that will surprise you. For example, a solution that everyone agreed to at the wrap-up meeting has changed into something very different, or points that were not an issue before suddenly become significant.

To help understand this, there are some things to keep in mind:

- **Things can look different when they are examined more closely.** An auditor found that batch manufacturing records contained too many errors and corrections. While it at first appeared that people needed more training on record-keeping and arithmetic, the problem was really that the forms were poorly designed and the electronic calculators were not reliable. If the response is different from what you expected, look for rationale as to why it was changed or talk to your contact for more information.
- **"Many roads lead to New York."** Since the audit, the auditee might have found a different, better way of reaching the goal. The solution they select has to work for them, or it won't work at all. For example, if the auditors found a deficiency in labeling of in-process equipment, rooms, etc. there are a number of methods for correcting the problem. Some are more "high tech" than others, but many can be suitable in a given operational/organizational environment. Figure 13.2 illustrates this.
- **"The auditors were wrong."** Disagreements on the facts can be reduced if the audit team confirms their observations during the data gathering phase of the audit. Also, the notes and working papers are useful. However, sometimes auditors *are* wrong. As you work through this, determine if the issue is about words (semantics) or is really an error.

Figure 13.2. Alternatives for identifying a manufacturing room.

Ways to identify a processing room with product name, code, and number are as follows:

- Small video monitor tied to manufacturing execution system giving specific information.
- Preprinted sign with identity information taped to room door.
- Identity information handwritten on small "white board" with appropriate markers.

Tracking Corrective Actions

Sometimes auditees will want to monitor the progress of corrective actions, especially if they are complex or will take time to implement. Figure 13.3 shows one such tracking system that auditors may find useful as they monitor the progress of the auditee.

Verifying Corrective Actions

When the auditee says that they have corrected the deficiency, what should you do? It depends on several factors, such as the trust the audit sponsor has with the auditee, the organizational relationship, and the seriousness of the observation. Usually "trust but verify" is adequate.

If high levels of trust exist between the auditors and auditee, simply following up at the next audit will be suitable for most situations. If the deficiency is serious, there may be other ways to monitor that it has been corrected, for instance, by requesting a copy of the SOP or the completed validation report. If trust levels are low, you may

Figure 13.3. Example of a system for tracking audit corrections.

Audit date: Oct. 13–16, 1997
Update report date: Oct. 25, 1997

Dept.	Audit Ref #	Commitment	Assigned to	Target	Done	QA OK
Micro lab	1.1	PM on autoclave	J. Wilson	Oct. 24/97	Oct. 23/97	Oct. 23/97
Dispensing	2.1	Update SOP #25.4	C. Currick	Oct. 24/97		
	2.2	Training on SOP #25.4	C. Vega	Oct. 31/97		

want to schedule a follow-up audit. Drawbacks to this are that trust levels may diminish even more and relationships could be strained.

Another way to verify a correction is to ask others to assist. For example, if you are working with a contractor and have a "man-in-the-plant," you can ask that person to look for specific things. Or, if other technical personnel periodically visit the site, you may want to ask them to follow up.

Verifications are more easily done if good communication exists and if the organization values sharing ideas and learning from one another. Management needs to understand the advantages and disadvantages of the different types of verification.

Closure

When the lead auditor feels that all issues have been or are being adequately addressed, he or she issues a *closure memo* to management. Those receiving this memo are typically the same addressees who received the audit report. A copy of the closure memo should also be sent to the auditee. If there are any open issues or things that must be verified in the next audit, they should be included here.

Archiving Working Papers

An "official" file should be identified in your auditing SOP as the place where all audit reports and working papers are kept. These should be controlled, confidential files kept for a designated period of time. Record retention time for all working papers should be for at least one or two audit cycles; reports, responses, and closure memos should be kept for 5–7 years.

As mentioned in the previous chapter, there may be legal considerations about copies of audit reports, working papers, and the length of time they are retained. Discuss this with the management and legal staff.

Milestones for Phase V: Follow-up and Closure

You know you have completed one audit cycle when you have:

- A satisfactory written response from the auditee explaining what they are going to do and improvments they are going to make;

- Confidence that the auditee will meet its commitments;
- Organized and archived all relevant audit reports and working papers; and
- Published a memo to management that the improvements are on target and that the audit has been closed out.

CHAPTER 14

Auditing Vendors and Contractors

Goals

- Identify several issues that are particularly important when auditing vendors and contractors.
- Discuss different audit approaches that may be appropriate while auditing vendors and contractors.
- Discuss auditing/response issues from the vendor's point of view.
- Discuss audit report issues.

Auditing vendors and contractors is one element in qualifying a *contract acceptor*. A contract acceptor is an organization that has agreed to provide information (e.g., an analytical laboratory), goods (components, active ingredients), or services (contract manufacturing or packaging). Usually, a firm will audit a vendor or contractor before the contract is signed, and then periodically audit the vendor or contractor during the relationship. In addition, if there is a significant problem, a drug or device firm may conduct a special audit in order to resolve the issues.

Note: Unless specified, the term *vendor* will include both suppliers and contractors. *Vendor*, as used here, will also be the same as *contract acceptor*.

As you audit a vendor, not only will you be examining their facilities and quality systems, but the vendor will be evaluating you and your colleagues. To them, *you* will represent your organization. The relationship between the organizations will be shaped by how well or poorly the audit goes.

Auditing a vendor may have one of several goals:

- For a new vendor—determining if they have the facilities, capabilities, and controls required to produce a product to the client's requirements.
- For a current vendor—evaluating if they are producing the product (or information) to your firm's standards and other (e.g., GMP) requirements.
- For a vendor that has problems—evaluating the significance of the problems, identifying corrective actions and plans, and determining whether customer/vendor relationship should continue.

As discussed in previous chapters, the audit approach(es) selected and used need to accomplish the audit goals.

The vendor will also have some goals *they* will want to accomplish during the audit, such as the following:

- Showing the best parts of their organization,
- Making a positive impression on the auditors,
- Discovering how they compare to the auditor's expectations,

- Obtaining intelligence as to how they compare to competitors and others in the industry,
- Learning as much as they can about the customer,
- Getting or keeping the contract,
- Not letting out confidential information, and
- Demonstrating their organization's technical and business expertise.

There may be other goals, spoken or unspoken, on the part of both the customer and vendor. Sometimes, these may be in conflict with each other, while at other times it is quite easy to accomplish a "win-win" solution for both parties.

There may be times when it is difficult to travel to a remote site or when a firm doesn't have the resources to make vendor visits. An alternative method is to conduct some or all of the audit by phone or fax. One approach is provided by Lyall and Anisfeld. (See listing in the Resources section.)

Special Preparation for Vendor Audits

The audit cycle for a vendor uses the same five phases (preparation, conducting the audit, analyzing audit data, reporting the results, and follow-up and closure) presented in earlier chapters; however, some of the tasks involved in each phase are slightly different from those that would be done for an internal audit. Some of these differences are discussed below.

Phase I: Preparation

If you are auditing a vendor for the first time, look for information about the vendor in databases or by networking with colleagues through professional and technical organizations. For vendors who may have been audited by the U.S. FDA, you may want to request copies of recent inspection reports, FDA Form 483 observations, and Warning Letters. This can be done directly through the FDA Freedom of Information Office or by using an intermediary source that can do this without identifying your firm.

As you prepare for the audit, you will want to be sure there is a contractual agreement between the vendor and your firm that provides for initial and periodic audits. This should be part of the general agreement your firm has with the supplier.

Having confidentiality agreements between both the vendor and the customer will allow the sharing of information and should facilitate your seeing the vendor's facilities. These agreements should be executed *before* the audit takes place so that legal groups from both sides have an opportunity to review the agreements. It is important that the agreement bind and protect both parties.

As you develop the list of requirements in the audit checklist, identify those that are of highest priority or are "show-stoppers." If the vendor provides most of its goods or services to other industries or is a new supplier to FDA-regulated firms, it may or may not have the systems and controls in place that are common in drug and device firms. In other words, know what is truly important to your organization.

It is also important to have an understanding of the vendor's strategic importance to your firm. That is, is this vendor one of many who wants to supply your needs or is there something unique about the vendor's products or services that your firm very much wants or needs to use?

One other aspect of audit requirements (which is also true with specifications) is that each requirement should have value to the customer. Why? Because many requirements cost money to implement and confirm through your testing. For example, it will cost more money to the vendor (and eventually the customer) to have components rinsed with Water for Injection (WFI) than with deionized water. If WFI is not needed, it shouldn't be required, as it would not be of any added value. In other words, you don't want to pay for quality that will not add value to your firm or your products.

In communicating with the vendor as you plan for the audit, don't focus strictly on compliance; the focus should be on quality, continual improvement, and meeting the needs of each of the parties.

As you create checklists for auditing either potential or current vendors, identify the key issues to your firm. They may include things such as:

- Component/product quality,
- On-time delivery,
- Overall capabilities,
- Reliability of laboratory testing results,
- Consistency of product,
- Their GMP systems,
- Their GMP compliance history with regulators,
- Responsiveness,
- Workforce stability,
- Technical service/support,
- Personnel availability/expertise,
- Cost of product/value, and
- Critical technology.

Gathering concerns about the current vendor and its products is useful as you prepare for the audit. In particular, you should contact people in purchasing, materials planning, receiving, and sampling/quality control and ask for their experiences. Also, talk to people who use the product, such as operators, technicians, and supervisors. You may discover that operators are much more sensitive to real-life product quality attributes than QC people, who test against the product specifications. For example, operators in a parenteral product formulations area saw a significant variation in the "clumpiness" of an active ingredient. They knew from experience that clumpy material would dissolve much more slowly than the powdered material. GMP auditors could ask about this issue as they conduct their audit of the vendor. Another source of information you should access are deviation reports. These may show problems your firm has experienced with the vendor's product(s).

Several weeks before the audit, the lead auditor should request that the vendor send materials to the audit team to be examined as part of the desk audit. Don't be overly concerned, however, if the vendor replies that they cannot accommodate this request; many vendors will not send drawings, SOP lists, master validation plans, etc. to any outside organization. If this is the case, ask that these materials be available to the audit team at the start of your visit.

The week before the audit, you should phone the vendor to confirm the time and goals of the audit. If there are particular areas or processes you want to see in operation, confirm that they will be accessible and running.

Certain GMP regulations, such as those written by the European Community and the Canadian HPB, have additional details regarding the relationship between customers (contract givers) and vendors (contract acceptors), in particular those vendors who are contract manufacturers. These are very useful reference documents to any firm working with vendors.

Phase II: Conducting the Vendor Audit

A vendor audit will typically start out with the opening meeting. In addition to the usual introductions, it is helpful for the first-time auditors of the facility to ask about the history of the organization, its quality program, and the firm's regulatory experience. This not only helps with building the relationship between the two groups, but can confirm the preliminary research the auditors have done. The auditors should note who is present at the opening meeting (including their titles) and their knowledge of the issues.

The audit should include an opportunity of seeing the facility in action—watching personnel assemble components, testing materials, and manufacturing the item(s) purchased by the customer. As you examine the vendor's quality system, its processes, equipment, and facilities, you may see activities that are different from your expectations—don't impose on the vendor your firm's specific solutions to problems. Also, before being critical, be sure your expectations are reasonable. For example, if you are auditing an active pharmaceutical ingredient manufacturer, it will typically look more like a chemical plant than a finished drug facility, particularly in the early steps of the process. GMPs will be applied slightly differently at the bulk drug facility as well.

As you proceed through the facility, you will be evaluating not only the equipment, operations, documents, and systems, but the personnel as well. One auditor, during the plant tour of a contract manufacturer, was being shown tablets coming from the tablet compression machine. The QC inspector hosting the tour intercepted a sample of tablets in her bare, unwashed hand as the tablets were being discharged from the machine into a plastic bin. After admiring the quality of the finished tablets, the auditor held his breath—what would the QC inspector do with the tablets in her hand? The answer, which ended up as a critical observation, was that the inspector simply placed the tablets in the plastic bin with the other, untouched tablets. This indicated to the auditor that personnel really didn't understand the basics of protecting the purity of the drug products.

Phase III: Analyzing Audit Data and Phase IV: The Audit Report

Both phases are considered together here because they depend on the audit sponsor's procedures and your relationship with the vendor. Vendor audit teams from some organizations are permitted to make decisions about vendors as they do their audits. Other firms require audit teams to make recommendations on whether or not to use a vendor to a corporate group, and the corporate group communicates its decision to the vendor.

Corrective actions that should or must be taken can become an issue with vendors, especially those who are trying to accommodate many different firms in the drug and medical device industry. If problems are observed, focus on the basic issue or why the item does not meet your requirements. Avoid telling the vendor what to do.

At other times, the vendor may ask for solutions to a problem. This can also be a difficult situation because if the ideas are misunderstood or are not implemented properly, they could come back and blame the audit team. If asked for a solution, give examples of different ways the issue can be corrected. If the vendor needs to make a significant investment in GMP changes, suggest they use an outside consulting firm as a resource that can more fully consider the changes and their implications.

Phase V: Follow-up and Closure

The points covered in Chapter 13 are valid for vendors as well, with the follow-up actions primarily dependent on the trust level between the firms and the seriousness of the deficiencies.

Conclusion

Auditing vendors is an important step in building quality into the drug or medical device products. It is also a way of creating an honest and communicative environment between supplier and customer, as well as minimizing surprises due to misunderstandings or changes. Like other types of audits, vendor auditing is an investment that can have tangible and intangible dividends.

CHAPTER 15

Being Audited

Goals

- Describe actions the auditee can take before, during, and after the audit.
- List items that should be addressed in a procedure concerning responding to an audit or regulatory inspection.
- Identify things to do and not to do during an audit or regulatory inspection.

Up to this point, our emphasis has been on equipping GMP auditors with the knowledge, tools, methods, and questions that they can use while preparing for and conducting a GMP audit. In this chapter, we are going to look at GMP audits from another perspective, that is, from the point of view of the auditee. We will follow the same five-phased approach, as shown in Figure 2.1. Unless specifically mentioned, the information below is applicable to internal audits, regulatory inspections, and other third-party audits (e.g., vendor or contractor).

183

Phase I: Preparation

If you are in the medical device or pharmaceutical industry, or a supplier to the industry, you *know* that you will be audited. You could be receiving an audit from

- Your firm's corporate quality assurance department (a first-party, internal audit);
- A customer (second-party audit);
- A consulting auditor representing a firm (third-party audit); or
- A regulatory agency.

These could be routine audits, such as a biannual visit to your firm by a customer or someone from corporate QA. Or, the audit could be prompted by a special situation, for instance, when your firm has submitted a drug or device application to a regulatory agency. Just as the auditors are preparing to visit you, you should be preparing for the audit.

Have a Procedure in Place

There should be a defined, well-understood process for greeting and hosting auditors, be they corporate associates or regulatory officials. While not required by GMP, having such an SOP is an industry practice. The procedure should include the following:

- Who is notified when an auditor arrives;
- Meeting and greeting the auditor in a reception area;
- Reviewing credentials and reasons for the inspection;
- Who should attend the kick-off meeting;
- How to respond if a team of auditors arrives;
- Accompanying the auditor(s) at all times;

- Where to place the auditor(s) when they are working (e.g., a well-controlled conference room);
- Retrieving, copying, and double-checking information supplied to the auditor;
- Taking notes on all auditor activities, questions,
  documents provided;
- Keeping management informed about the progress
  of the audit;
- Expectations about proper, ethical behavior on the
  part of the firm and the auditor;
- Position on providing coffee and lunch to the auditors;
- Position on the taking of photos, videos, record-
  ings, etc.; and
- Who should attend the close-out meeting.

Evaluate the Areas and Systems

Evaluation may mean conducting a preparatory audit (see
Chapter 5) of the area, systems, or products that are likely
to be audited. To assess where you really are, you want
a thorough, tough auditor. This is not a time for self-
congratulations and giving everyone a warm feeling; you
want to understand your vulnerabilities and address them.

A key point in self-evaluation: timing. The earlier you
start this activity, the more time you will have and the
more issues you can correct.

Make Sure You Are Keeping
Your Commitments

If your firm had an audit or regulatory inspection several
years ago and management made commitments to the
organization or agency, be sure you are still keeping them.
Most auditors look at reports from the previous two or

three audit cycles to see if corrections that were promised are still being kept. This is important for customer audits as well: you want to show them you do what you say. If the auditors see you did not fulfill your commitments, your organization will lose a great deal of credibility and trust.

Address the Issues Strategically

If the evaluation has determined there are a number of problems, you may not have the time, people, or resources to correct them all and have them working perfectly by the time the auditors arrive. You need to determine which issues are show-stoppers, that is, what deficiencies will most likely cause you to fail the audit. You should know these by having someone in your firm (or an outside consultant) understand the current trends important to the agency or client and what they are expecting.

For example, you are developing a new product and the U.S. FDA will be conducting a Pre-Approval Inspection (PAI). This will be your firm's first marketed product, and it is not validated. You should know what the FDA is looking for in terms of validation. As of this writing, they will not require completed process validation of the product to recommend an approval; they will want the validation done before the product is marketed. During their PAI, the FDA will want to see, among other things, if utilities (e.g., water, HVAC) have been validated and if there are validation plans and protocols in place. In other words, separate out the "must haves" from the "nice-to-haves." Base this on current information, not just something someone read in a trade magazine a few years ago.

Look For the Root Causes

Many of the problems you find could be related to the same root cause. For example, maybe deviations are not being completed in a timely way because no one is keeping track of them. Yes, creating a deviation log will be a window on what is going on in your organization, but more importantly, it gives management a tool to monitor and improve things. If you do not have time to analyze a problem, understand where quick-fixes are being used and plan to go back and properly correct them. As you make changes, do not bypass change control. Changes need to be evaluated and documented.

Know the Weaknesses and Develop a Plan to Improve Them

If there are some deficiencies that you cannot properly correct in the amount of time available, at least develop a plan and timeline to address the problems. If the auditors discover the problems, you will be in a much better position to say that you have found the issues yourselves and have a plan mapped out. What if the auditors do not find the problems—should you proceed and make the corrections? Yes—you should have proven to yourselves that there are quality and business reasons for doing it, not just compliance reasons.

Ask for the Audit Plan and Other Information

As discussed in Chapter 8, auditors should have a plan that is available to the auditee. In some cases, they will also provide the auditee with a checklist of items they will be looking at. The more information you can get from the auditor, the better prepared you can be. Do not expect, however, that you can "cram" and get everything perfect in a short period of time. Quality systems cannot be created overnight and then be expected to work. Good auditors will be able to see through most last minute preparations.

Find Out Who the Auditor Will Be

If you are expecting a corporate audit team or a consultant, you will probably be able to find out who it is and gather some intelligence about him or her. If you are receiving an FDA Pre-Approval Inspection, you may be able to discover who will be arriving. If you can find out this information, network with colleagues to get an idea of their experiences with the person. If you are in a large corporation, talk to other sites that may have had experience with the auditor. (One firm is known to keep an informal database on various internal and external auditors, including areas of expertise and what they like to eat.)

For example, a European firm discovered that one particular FDA inspector was going to conduct their Pre-Approval Inspection for a sterile product. The agency official was an expert in aseptic processing, water system monitoring, and computer validation, but was known not to spend a lot of time in the laboratory. In preparing for the inspection, they covered all areas, but put additional emphasis on particular topics they thought would be of most interest to the inspector. They were right.

One cautionary note: if you gather background information like this, *do not rely too heavily on it*, as auditors also keep up with trends, issues, and regulatory mandates.

If You Are Doing Something "Different" Be Prepared to Justify It

In many of the GMPs or guidelines put out by regulatory agencies, they make a comments such as:

> During establishment inspections, the Health Protection Branch will use this publication [4th Edition on GMPs] as a guide in judging compliance. . .Alternative

means of compliance. . .also will be considered at such times (HPB 1996).

and

> This inspection guide is designed to establish inspection consistency and uniformity by discussing practices that have been found acceptable (or unacceptable). At the same time, one must recognize that for cleaning validation, as with validation of other processes, there may be more than one way to validate a process. In the end, the test of any validated process is whether scientific data shows that the system consistently performs as expected and produces a result that consistently meets predetermined specifications (FDA 1993).

In other words, customers or regulatory agencies may have one preferred way of doing something. If you do it that way, you may need only a small amount of documentation to justify your position. On the other hand, if you are doing something very different from industry practice or from what regulatory inspectors expect, you will need solid, scientific evidence to show that what you are doing is as good or better than what is typically done. The typical questions that will be asked include the following:

- Why are you doing it this way?
- Why aren't you doing it as suggested in the guide/ guideline or the way that most of industry is doing it?
- How do you know it works?
- Where are the data that support doing it this way?

Identify the Experts

As you are preparing for the audit, think about questions, documents, systems, products, and processes that might be targeted. Then, find the people who can most clearly and knowledgeably talk about them. Do not expect the audit coordinator or a member of upper management to know all the details. It is better to bring the right person in than to attempt to answer any questions and potentially confuse the auditor.

Some firms develop a list of their expert personnel and topics they can address. One firm gives these experts pagers to wear during an audit so they are always easily accessible to the audit coordinator.

Prepare Your People

The reactions of people in your organization, when they hear that an audit will be happening, will range from panic to blasé weariness. Either of these extremes can cause potential problems. At the very least, all personnel should know:

- Why they are being audited,
- Consequences of the audit,
- What auditors and agency inspectors can and cannot do,
- What auditors expect,
- How to respond to questions from an auditor,
- What you should say if you do not know an answer, and
- The importance of telling the truth at all times.

Many firms prepare their personnel with a training program or a mock inspectional activity before the auditors arrive. These can significantly reduce the apprehension of all personnel. (See the Resources section for an example of such a training course.)

Phase II: The Audit

The auditors should have a plan of what they want to do and specific things they want to look at. If they didn't communicate these with the firm early on, auditors should do this at the opening meeting. At the opening meeting, let the auditors know what *your* expectations are, for example, if you want an informal briefing session at the end of each day or your process for the auditors' requesting copies of documents.

It is customary to give the audit team a private conference room from which they will work. It should be equipped with a phone. It should not have any other corporate documents like SOPs, phone directories, policy manuals, and the like. (If you are designing a facility, consider preparing such a room off the main lobby that has easy, yet controlled access to restrooms, water, etc., but will not allow auditors to "wander" into laboratories or manufacturing rooms.)

If the auditor wants to talk with people, try to bring the people to the auditor in the conference room. However, most auditors will want to see the person in his or her work environment. If this happens, alert the people in advance (or that you are on your way) and try to have a clean area available for the conversation. Otherwise, the auditor may inquire about documents, objects, and other items he or she observes. Many auditors have developed the uncanny skill of reading things upside down.

As you take an auditor through a facility, be very careful about steering him or her. Most auditors can sense when they are being led and controlled and will react negatively to this.

One of the most important things in an audit is to supply information and documents that are requested as quickly as possible. *Readily available* and *readily accessible* mean just that. Auditors, in particular regulatory inspectors, have learned that the longer they wait, the more suspect the documents become.

Not only do you need to have the documents, but you also need to have a systematic way of storing and obtaining them. If the auditor is requesting batch records of a product made three years ago, and they are kept offsite on microfilm, explain it to the auditor up front so he or she knows what to expect in terms of waiting time.

Each auditor has his or her own definition of *readily available*, but a rule of thumb is:

If the document or record is	it should be available within
an SOP or work instructions personnel should be able to normally use	5 minutes
a department record kept in the immediate area (e.g., room cleaning log completed 4 months ago)	15 minutes
records kept at the site but in another department	1–3 hours depending on age of the document and how it is archived
records kept at another location	2–4 hours by fax depending on age of the document and how it is archived; originals, if required, by next day

While these are not hard and fast rules, the rationale of the auditor would be that if you or your people are expected to use these documents in making decisions or performing a task, they should be convenient and quickly accessible.

It needs to be stressed to all personnel that answers to questions and documents presented need to be truthful.

If a firm's integrity is questioned, it could mean a serious blemish on its reputation in the industry or, more seriously, civil or criminal action.

Some other "dos and don'ts" about what to do during an audit:

- Understand the question before answering it.
- Keep your response as short as possible; use "yes" or "no" whenever you can, but, at the same time, answer so that the auditor will be satisfied.
- Bring a specific document when requested, not a full file or binder.
- Do not volunteer information.
- Answer with facts.
- Do not talk and work at the same time.
- If you do not know something, do not guess.
- Directly refer to procedures and other documents, especially when it comes to times, rates, specifications, etc.
- Do not try to be overly helpful.
- Answer questions and behave professionally, courteously, and carefully.
- Be confident of yourself—you know more about what you do in your position than the auditor knows.
- Always tell the truth.

If the auditor finds a problem, determine if it is easily correctable. If it is, make the necessary changes, documenting them and using the change control process if required. Particularly with FDA inspectors, if a change has been made, show them the correction and attempt to not have it written on a Form FDA 483. It is easier to prevent it from being written than getting an inspector to eliminate an observation once it is written on the FDA 483.

When working with internal auditors or consultants who find a problem, talk to them about options for correcting it. They may have some novel ideas of how to solve an issue that are simple and effective. (Most regulatory agency personnel are reluctant to provide any suggestions, as it may be misinterpreted or not implemented as intended; or they may be wrong.)

Phase III: Analysis

This activity is primarily done by the auditors; there is little for the auditee to do.

Phase IV: Report

When the auditors present their preliminary report at the close-out meeting of the audit, ask for as many specifics as possible, such as number of occurrences and where they saw the problem. This should confirm the information from the audit contact(s), who worked directly with the auditor(s). You are not doing this to make life difficult for the auditors but to collect information to use in your corrective action plan.

Between the time that you hear the preliminary findings and the time you receive the final report, begin formulating the options available to you. If there are some issues that you feel you can defend, start collecting the data and preparing your rationale. The more time you have for considering options, the better your response can be. Some firms have made commitments in the heat of an audit and then realized that there were other, better options available to them.

When the report arrives, review it carefully to be sure the observations and their significance have not changed.

If there are any questions or if there is anything that is unclear, call the lead auditor immediately and discuss these with him or her.

In the response to the audit (or regulatory inspection observations), you will want to provide enough detail to the auditor so he or she can determine that the corrective actions will be appropriate, adequate, and address the root problem. Do not limit your response only to the particular item or product. Consider if the correction should be extended to similar operations or products.

Sometimes a solution includes both short-term and longer-term activities. Be specific about what will be done when and who is responsible. If several options for a long-term solution are being evaluated, describe what you are doing and when results will be available.

The more details you can provide the auditor in the response, the more confidence he or she will have that you know and understand what you are doing. Vague, quick answers will not satisfy most auditors and can precipitate a follow-up audit.

What if you disagree with the audit report? For example, what if the audit team criticizes a practice you feel is adequate and meets GMP? First, understand the essence of the auditors' observation. If they did not provide adequate insight into the issue, ask them for more information. Which of the GMP values (product safety, identity, strength, purity, or quality) do they feel is at risk? Is your approach slightly different than they had expected or completely at odds with industry practice that they have observed? (The latter will be more difficult to justify.)

Second, determine how *your* current approach accomplishes the GMP values of safety, identity, strength, purity, and quality. How do you know it is effective and in control? What data or records do you have to prove it? Talk to the lead auditor and share your concerns. Work towards a common understanding and solution.

Phase V: Following up

It is critical that commitments you have made are accomplished in a timely way. Some firms use a timeline/checklist of their commitments and review them at high-level staff meetings. One of the worst things that can happen is to be reaudited and not have completed the promises that were made.

If situations change while you are implementing a corrective action, such as a piece of equipment is delayed or business conditions have prevented your firm from accomplishing a commitment, talk with the auditor. You will be doing at least two things with this. First, you will be communicating so as to avoid surprises. Second, you may be able to either work out a new timeline or find a different way of solving the problem based on the current conditions.

Conclusion

Being audited is not an event that firms look forward to, especially when the outcome is critical to the life of the company. At the same time, if a company is reasonably prepared and has basic quality systems in place, audits should not be feared. They are a valuable learning experience, an assessment tool, and an essential element for continual improvement.

REFERENCES AND READINGS

ANSI/ISO/ASQC. 1994. *Guidelines for Auditing Quality Systems.* ANSI/ISO/ASQC 10011-1994. Milwaukee, WI: ASQC Press

Brassard, M., and D. Ritter. 1994. *The Memory Jogger™ II.* Methuen, MA: GOAL/QPC.

EC. 1992. *The Rules Governing Medical Products in the European Community, Vol. IV, Good Manufacturing Practice for Medicinal Products.* Luxembourg: Office for Official Publications of the European Community.

FDA. 1992. *Foreign Inspections Guide.* Rockville, MD: Food and Drug Administration.

FDA. 1993. *Cleaning Validation.* Rockville, MD: Food and Drug Administration.

FDA. 1996a. Title 21 Part 820—Quality System Regulation for Medical Devices. *Federal Register,* Vol. 61 No. 195, October 7, 1996; pp. 52654–52662.

FDA. 1996b. *Investigation Operations Manual.* Rockville, MD: Food and Drug Administration; p. 239.

FDA. 1996c. Title 21 Part 820—Quality System Regulation for Medical Devices. *Federal Register*, Vol. 61 No. 195, October 7, 1996; p. 52614, comment 55.

GMP Letter. 1996. August, Issue No. 199; p. 2.

The Gold Sheet. 1996. FDA Reviewing Certification Options. April, Vol. 30, #4.

HPB. 1996. *Good Manufacturing Practice Guidelines*, 4th ed. Ottawa, Ontario, Canada: Health Protection Branch Drugs Directorate.

Lamprecht, James L. 1992. *ISO 9000: Preparing for Registration*. Milwaukee, WI: ASQC Press; p. 136.

Mager, R.F., and P. Pipe. 1984. *Analyzing Performance Problems*, 2nd ed. Belmont, CA: Lake Publishing Co.

Mehrabian, Albert. 1981. *Silent Messages*. Belmont, CA: Wadsworth Publishing Co.

Mills, Charles A. 1989. *The Quality Audit: A Management Evaluation Tool*. Milwaukee, WI: ASQC Press; pp. 169–180.

Vesper, James L. 1993. *Training for the Healthcare Manufacturing Industries*. Buffalo Grove, IL: Interpharm Press; p. 15.

Weightman, R. T. 1994. How to Select a Registrar. In *The ISO 9000 Handbook*, 2nd edition. Ed. Robert W. Peach. Fairfax, VA: CEEM Information Services; p. 127.

Wolin, Alfred M. 1993. U.S. v. Barr Laboratories, Inc. Newark, NJ: U.S. District Court, Civil Action 92–1744.

American Society for Quality Control Code of Ethics

To uphold and advance the honor and dignity of the profession, and in keeping with high standards of ethical conduct, I acknowledge that I will:

Fundamental Principles

I. Be honest and impartial; will serve with devotion my employer, my clients, and the public.
II. Strive to increase the competence and prestige of the profession.
III. Use my knowledge and skill for the advancement of human welfare and in promoting the safety and reliability of products for public use.
IV. Earnestly endeavor to aid the work of the Society.

Relations With the Public

1.1 Do whatever I can to promote the reliability and safety of all products that come within my jurisdiction.

1.2 Endeavor to extend public knowledge to the work of the Society and its members that relates to the public welfare.

1.3 Be dignified and modest in explaining my work and merit.

1.4 Preface any public statements that I may issue by clearly indicating on whose behalf they are made.

Relations With Employers and Clients

2.1 Act in professional matters as a faithful agent or trustee for each employer or client.

2.2 Inform each client or employer of any business connections, interests, or affiliations that might influence my judgment or impair the equitable character of my service.

2.3 Indicate to my employer or client the adverse consequences to be expected if my professional judgment is overruled.

2.4 Not disclose information concerning the business affairs or technical processes of any present or former employer or client without his or her consent.

2.5 Not accept compensation from more than one party for the same service without the consent of all parties. If employed, I will engage in supplementary employment of consulting practice only with the consent of my employer.

Relations With Peers

3.1 Take care that credit for the work of others is given to those whom it is due.

3.2 Endeavor to aid the professional development and advancement of those in my employ or under my supervision.

3.3 Not compete unfairly with others; will extend my friendship and confidence to all associates and those with whom I have business relations.

Source: Certified Quality Auditor Certification Program, June 1991. Reprinted with permission from American Society for Quality Control.

In Their Own Words: GMP and Quality Auditors Talk About Auditing

Author's Note: In the fall of 1996, a request was made to readers of an on-line discussion group, asking those with auditing experience to complete a questionnaire. The following responses were received. Only minor editing changes were made to improve the readability. Thanks to all of those who participated in this!

Questionnaire #1

Name: Michael H. Anisfeld
Company: Interpharm Consulting
Address: Buffalo Grove, Illinois

How long have you been involved in GMP auditing?
Since 1980.

What was your original background?
MS in Industrial Pharmacy with work experience in production (3 years), Materials Management (1 year), Quality Control (2 years), Development (2 years), and Quality Assurance (2 years)—all before I started in GMP auditing.

What are the most important characteristics of a successful auditor?
The ability to look beyond the obvious symptoms, and look at the root causes of the problem—and then the ability to effect change.

What audit do you remember most vividly and why?
An audit I performed for the United Nations in China. Having to rely on a nontechnical translator who felt that whatever translation gave an answer that sounded good was the one to give. When you are spending three weeks in the far reaches of China, you desperately need the interpreter.

From your perspective, how do GMP audits add value?
If the audit reveals a weakness in the system that could have an impact on patient safety (death or injury), it has added value.

What is your greatest fear as an auditor?
Missing a vital area due to lack of time to cover everything, or lack of concentration due to jet lag.

What is the quality you most want a co-auditor to have?
Perseverance and the ability to kept following the thread until the root causes of noncompliance are discovered.

What is one of the most important questions to ask during a GMP audit?
Who pays the salary and provides the performance review

of the most senior "quality" person on-site. This one question is the quickest way to reveal any conflicts of interest between production and quality groups.

What advice do you have for a person starting out as a GMP auditor?
Do not speak for more than 20–25 percent of an audit. Teach yourself to listen for the most nonverbal as well as verbal clues.

After completing an audit, what is your favorite thing to do?
Helping effect change and correcting the issues.

Questionnaire #2

Name: Tony Cundell
Company: (information not available)
Address: (information not available)

How long have you been involved in GMP auditing?
I audited blood banks for 18 months.

What was your original background?
I am a Ph.D. Microbiologist.

What are the most important characteristics of a successful auditor?
Desire to find the weak points in an organization. Perhaps ego!

What audit do you remember most vividly and why?
Lab manager was physically sick and tendered her resignation after finding that an HIV-positive unit was released to inventory.

From your perspective, how do GMP audits add value?
Inform upper management on the level of performance of their organization.

What is your greatest fear as an auditor?
1. Being beat up in the parking lot, and 2. Hurting people's feelings.

What is the quality you most want a co-auditor to have?
I always worked alone.

What is one of the most important questions to ask during a GMP audit?
I have read your SOP, but can you tell me how you do this operation?

What advice do you have for a person starting out as a GMP auditor?
Know the business you are going to audit and decide what are the one or two most important issues/systems that you will investigate.

After completing an audit, what is your favorite thing to do?
Issuing the audit report and helping with the corrective actions.

Questionnaire #3

Name: Tony Fugate
                Manager, Quality Assurance and Validations
Company: Pharmacia and Upjohn Pharmacia Hepar, Inc.
Address: (information not available)

How long have you been involved in GMP auditing?
Almost 10 years.

What was your original background?
I have a BA in History and Mathematics. Out of college, I immediately began working for the pharmaceutical industry.

What are the most important characteristics of a successful auditor?
Listening, observing details, judging character of auditee, being able to get along with others, not letting your ego show or get bruised, observing and reporting only the facts as compared with some standard.

What audit do you remember most vividly and why?
An audit of a testing laboratory in which I noted a log entry not made for the previous day. When I returned to look at it again, the log was completed.

From your perspective, how do GMP audits add value?
Compliance with regulations. In general, it saves time and cost.

What is your greatest fear as an auditor?
That I will not observe something of significance.

What is the quality you most want a co-auditor to have?
Same as above.

What is one of the most important questions to ask during a GMP audit?
There is no one important question. As a general rule, I observe any systemic problems with generic questions.

What advice do you have for a person starting out as a GMP auditor?
Learn by observing, and don't just limit yourself to product or audit knowledge. Obtain training on dealing with people and how to deal with problem people. This was my most valuable training.

After completing an audit, what is your favorite thing to do?
Complete the audit report.

Questionnaire #4

Name: Douglas B. Hecker
Company: MidWest Research Institute
Address: Kansas City, MO

How long have you been involved in GMP auditing?
Since 1991.

What was your original background?
Analytical chemistry, methods development/validation, and process optimization/validation.

What are the most important characteristics of a successful auditor?
Ability to organize, prioritize, and socialize. Additionally, being able to "read" the auditee (i.e., know when they are just feeding you a line).

What audit do you remember most vividly and why?
There are several. 1) A company in Padova, Italy (BPC manufacturer). When asked about how raw materials are organized and FIFO practice, stated it was all done by computer; however, when I asked to look at it, they informed me "No, we do not consider this important nor part of a

regulatory requirement"! 2) At a BPC in Rome, I observed a Batch Record that required a sample to be taken for residual solvents testing near the end of the drying process. The operator did not document the sample being taken, nor the results from the laboratory and proceeded to drum the finished product. The company's QA did not understand the problem with this.

From your perspective, how do GMP audits add value?
A GMP audit tests/evaluates your quality system's robustness. Each company should view an audit as a tool for training their new/current employees because the auditor is getting a quick view (learning and understanding it) of the company's system that is similar to what the company would do to a new employee. If the company's system is difficult for an auditor, just imagine what it is like for new/green employees.

What is your greatest fear as an auditor?
Not having enough time to complete a thorough audit—and missing something.

What is the quality you most want a co-auditor to have?
Interpersonal skills.

What is one of the most important questions to ask during a GMP audit?
Can I see your company's response to your last two FDA audits? Some companies provide great responses; however, what they say and what they do are often quite different.

What advice do you have for a person starting out as a GMP auditor?
Get experience in as many different areas as you can prior to becoming an auditor.

After completing an audit, what is your favorite thing to do?
Evaluate how good the audit was; did I meet my goals/expectations for the audit? Did I find a weakness in the company's quality system that could be corrected and would benefit that company as well as the company I worked for, and ultimately the public? Did the auditee and I work well together for the common goals, compliance? Also, kick back and enjoy the flight home.

Questionnaire #5

Name: Marion Weinreb
Company: Michaels and Weinreb Consulting Services
Address: San Francisco, California

How long have you been involved in GMP auditing?
I have been auditing for approximately 18 of my 20 years of professional experience.

What was your original background?
BA Biology, MS Pharmaceutical Marketing, with 20 years in the healthcare industry.

What are the most important characteristics of a successful auditor?
Excellent verbal, listening, interpersonal and written skills, ability to contain ego, ability to maintain objectivity during audit process, friendly and amicable personality, ability to ask the right questions even if the technical knowledge is not there.

What audit do you remember most vividly and why?
An audit of an unethical sterilization facility where I observed

fraudulent data entries. I thought they were going to physically harm me and I was alone in the facility.

From your perspective, how do GMP audits add value?
GMP audits help isolate system, process, product, and specific problems that stand in the way of maintaining high quality. Audits are educational and eye opening for the participants as well as the auditor.

What is your greatest fear as an auditor?
I fear that I will not uncover a system's problem that will later result in a recall or a significant, adverse quality issue.

What is the quality you most want a co-auditor to have?
Same as the auditor identified above. Plus the ability to get along with the author.

What is one of the most important questions to ask during a GMP audit?
I don't think there is one most important question that can be asked. If I have to answer that, I would ask the auditee: Does management support the quality program in place at their company and has management signed a quality policy.

What advice do you have for a person starting out as a GMP auditor?
Read as many articles as you can on auditing. Keep up with the literature on the areas that you will be auditing. Take as many courses as you can. Go on many audits with experienced auditors. Talk about the audit with them as it is happening.

After completing an audit, what is your favorite thing to do?
Go home and relax.

Questionnaire #6

Name: Ian E. Williams
Company: CP Pharmaceuticals
Address: Wexham, United Kingdom

How long have you been involved in GMP auditing?
16 years.

What was your original background?
Degree and professional qualification in Pharmacy, post graduate studies in Pharmaceutical Engineering Science, combined with lecturing and teaching. Four years pharmaceutical process development, four years as Medicines Inspector with UK Department of Health. Now QA Manager of pharmaceutical company in the UK.

What are the most important characteristics of a successful auditor?
Not to believe that you are the technical expert in the operations of the audited company, but the ability to see what others can't and to discern with technical competence the true significance of audit findings. Little is worse than an auditor without practical experience of real manufacturing life making major issues of finding things that have no product or process quality significance. Confucius said "Better a diamond with a flaw than a pebble without." Equally important is the ability to relate personally with the senior people within the audited company and, with rare (bad!) exceptions, to recognize that they share the same GMP goals and values as the auditor. Jackboots are not suitable clothing for auditors. Finally, when deficiencies are identified, to be able to unearth the real cause.

What audit do you remember most vividly and why?
A company in Italy where the combined production Manager/

QA Microbiologist, when questioned on the suitability of a garment sterilizing cycle of "about 110° C for about half an hour," replied "Well, when the garments come out they look sterile to me, and I'm a microbiologist!" He was also the microbiologist who incubated his yeasts and molds on bacterial agar at 37° C and passed every batch.

From your perspective, how do GMP audits add value?
In the context of auditing your own suppliers, there are four ways: 1) Building relationships and technical respect between the companies; 2) Identifying quality weaknesses or strengths that can enable the customer to design incoming QC intelligently; 3) Seeking to remedy GMP weaknesses—that will benefit both parties; and 4) Establishing credibility for your own company as one that is sufficiently serious about supply quality to carry out the audit.

What is your greatest fear as an auditor?
The biggest challenge is to understand the business, technology, and quality system of the audited company. To fail in those areas is to fail the audited company and your own. That's the biggest fear, to do a bad job, in other words. Tiredness can lead to lack of concentration. I have been audited by an auditor who kept falling asleep! I've never been that bad, or at least I don't think so! One fear I do not have is finding no deficiencies, unless it is because of failure to do a good job. It rarely happens, because there are no companies without weaknesses, but because of the "snapshot" randomness of an audit process, the systems should be robust enough to withstand sensible auditing. After all, isn't that our target in our own companies?

What is the quality you most want a co-auditor to have?
The ability to think laterally, to advise when I'm going down dead-ends or off at tangents, and to observe with

different eyes the things I'm looking at, or even more the things I'm looking at that I'm missing!

What is one of the most important questions to ask during a GMP audit?

Not sure about that, but the most important thing to observe is who answers the questions. In whom does the knowledge of the systems reside? I get worried when all the questions are answered by the quality man, but then I get worried when he answers none of them. I want to see evidence of quality ownership in every area I look at.

What advice do you have for a person starting out as a GMP auditor?

Learn to listen, learn to think laterally, and learn when to take things at face value and when not to! Best advice is to get out there and do it. If possible, get trained on the job with an experienced auditor; if not, start auditing "good" companies where, if you make a mess of it, the consequences are unlikely to be great, but where there is plenty of experience to be gained in seeing effective quality systems of different types at work.

After completing an audit, what is your favorite thing to do?

Get home. Nothing beats that after a day or week away. Ever!

—————————— APPENDIX 3

Resources

Those involved in GMP auditing may find the documents and information sources listed below valuable.

Audit Checklist and Questions

Audit by Mail: Time and Cost-Effective GMP Audit Tools
John Lyall and Michael Anisfeld
Interpharm Press, 1995.
Ph: +1 (847) 459-8480   Fx: +1 (847) 459-6644

Compliance Auditing for Pharmaceutical Manufacturers
Karen Ginsbury and Gil Bismuth
Interpharm Press, 1994.
Ph: +1 (847) 459-8480   Fx: +1 (847) 459-6644

GMP/ISO 9000 Quality Audit Manual for Healthcare Manufacturers and Their Suppliers, 5th ed.
Len Steinborn
Interpharm Press, 1998.
Ph: +1 (847) 459-8480   Fx: +1 (847) 459-6644

International Drug GMPs, 4th ed.
Interpharm Press, 1993.
Ph: +1 (847) 459-8480   Fx: +1 (847) 459-6644

GMP Training Materials

Preparing for a Regulatory Inspection (courseware with video, instructor's guide)
LearningPlus, 1996.
Ph: +1 (716) 442-0170 Fx: +1 (716) 442-0177
Internet: http://www.learningplus.com

Internet-Based Resources

These site addresses (URLs) were valid at the time of publication. Using a Web-based search engine (e.g., Yahoo or AltaVista) is a method of finding other current sites.

Canadian HPB website: http://www.hwc.ca/hpb
U.S. FDA website: http://www.fda.gov
Regulatory Affairs website: http://www.medmarket.com/
    tenants/rainfo/master.htm

Selected Technical Articles on Particular Areas

GMP Quality System

"From Audit to Process Assessment—The More Effective Approach," J. Nally, et al. *Pharmaceutical Technology* (19:9) Sept. 1995, pp. 128–140.

"Measuring Quality Performance," R. Bishara, E. Kaminski, et al. *Pharmaceutical Technology* (18:3) March 1994, pp. 140–152.

Deviations

"OSS Laboratory Results in the Production of Pharmaceuticals," D. Barr and G. Dolecek. *Pharmaceutical Technology* (20:4) April 1996, pp. 54–62.

Validation and Change Control

"FDA Guide to Validation." (Available on the U.S. FDA website.)

"The Validation Story . . . ," R. Tetzlaff, et al. *Pharmaceutical Technology* (17:3) March 1993, pp. 100-112.

Data Quality and Integrity

"Validation Issues for New Drug Development: Part III, Systematic Audit Techniques," R. Tetzlaff. *Pharmaceutical Technology* (17:1) Jan. 1993, pp. 80–88.

Laboratory

"FDA Guide to Inspection of QC Laboratories." (Available on the U.S. FDA website.)

"FDA Guide to Inspections of Microbiological Pharmaceutical QC Laboratories," July 1993. (Available on the U.S. FDA website.)

Supplier, Contractor Requirements

"FDA Guidance for Industry: Manufacture, Processing, or Holding of Active Pharmaceutical Ingredients, Discussion Draft," August 1996. (Available on the U.S. FDA website.)

"GMP Inspections of Drug Substance Manufacturers," H. Avallone, *Pharmaceutical Technology* (16:6) June 1992, pp. 46–55.

"GMP Issues and Documentation for Bulk Pharmaceutical Contract Manufacturers," M. Valazza. *Pharmaceutical Technology* (16:1) Nov. 1992, pp. 48–51.

Computer System

"Computer System Validation: Auditing Computer Systems for Quality," G. Grigonis, Jr., and M. Wyrick. *Pharmaceutical Technology* (18:9) Sept. 1994, pp. 48–58.

Product—Investigational

"Documentation Requirements for PAIs," J. Lee. *Pharmaceutical Technology* (17:3) March 1993, pp. 154–164.

"FDA Pre-Approval Inspection Program Guideline." (Available on the U.S. FDA website.)

Product—Commercial

"FDA Guide to Post-Approval Inspections." (Available on the U.S. FDA website.)

JAMES L. VESPER

James L. Vesper designs and develops instructional courses and workshops for the pharmaceutical and medical device industries. He established and is president of the firm LearningPlus, Inc., and has had more than 16 years' experience in the pharmaceutical industry.

Mr. Vesper worked eleven years at Eli Lilly and Company, Indianapolis, Indiana. His first assignment was as corporate industrial hygienist, followed by three years in Corporate Quality Assurance. There, he was responsible for issues concerning the manufacture and testing of parenteral products made at Lilly facilities and third-parties worldwide. His last assignment was as Project Leader of GMP (Good Manufacturing Practice) Education and Instruction, establishing the department and its mission.

Since 1991, Mr. Vesper has been creating innovative instructional products for the pharmaceutical and healthcare industries using selected video and computer technologies as effective and efficient delivery media. Working as consultants with a wide variety of clients, his firm creates integrated curricula for personnel and customized training courses targeted to particular needs. He speaks and writes for various international technical and professional organizations, including ISPE, GMP TEA, PDA, Pharmaceutical Sciences Group, and PharmTech.

In addition to traveling extensively and working on-site in Singapore, Latin America, India, France, England, Puerto Rico, and Italy, Mr. Vesper has been on the faculty of Indiana University teaching courses in Industrial Hygiene and guest lecturing at other universities. He has appeared on stage in the Metropolitan Opera's production of *Aïda* at New York's Lincoln Center in a supernumerary role.

Mr. Vesper earned a B.S. in Biology from Wheaton (Illinois) College and a Masters of Public Health from the University of Michigan School of Public Health in Ann Arbor. He is a member of ASTD and PDA.

James L. Vesper
LearningPlus, Inc.
1140 Highland Avenue
Rochester, NY 14620 USA
Phone: (716) 442-0170
Fax: (716) 442-0177
e-mail: jvesper@learningplus.com
website: http://www.learningplus.com

INDEX

A

Accurate (records) 56
Action plan 91, 105, 106,
   109, 114
Active pharmaceutical ingredient 179
Active pharmaceutical inter-
   mediates 66
American Society for Quality
   Control 3, 4, 79
Analysis 16, 122, 133-142,
   180, 194
Areas of excellence 122, 149,
   153, 156, 159
Areas to audit 92, 93
Aseptic/parenteral manufacturing 74
ASQC. *See* American Society
   for Quality Control
Audit 1-2. *See also* GMP audit
  approaches 49-52, 90, 93,
    101, 116
  cycle 14, 17
  documents 164, 165
  executive summary 145,
    146, 149
  goal 91-93, 113
  kick-off 112-114
  limitations 122
  observations 156
  plan 15, 89, 91-107, 109,
    151, 155, 187
  procedures 34

  process 14-17
  report 91, 143-162, 164,
    180, 194, 195
  scheduling 90
  summary letter 146-148
  team 42, 85, 90, 91, 95,
    96, 105, 110, 134
  types 8-10, 39-47, 90, 111
Auditee 8, 87, 89-91, 95,
   101, 104, 106, 110, 165-168
Auditing skills 82
Auditor-in-training 86
Auditors 5, 6, 8, 42, 77, 86,
   184, 185
Avoiding problems 7

B

Benchmarking audit 44
Benchmark of performance
  7, 100
Benefits of audits 7, 8
Bias 128
Binary standard 97
Bluff 131

C

Calibration 60, 68
Calibration, equipment 64,
   117
Canadian Drug GMPs 79
Canadian Health Protection
   Branch (HPB) 3, 27, 46,
   47, 178

Canadian requirements 27-29
Cause-and-effect chart 140-142
Certification 35
Certification audit 46
CFR. *See* 21 CFR
cGMPs for Finished Pharma-
      ceuticals 20, 21
cGMPs for Medical Devices:
      Quality System Regula-
      tion 21-27
Change control 59, 69
Checklists 91, 98-103, 112,
      114, 138, 176, 177, 187
Cleaning practices 59
Clear (records) 56
Client. *See* Sponsor
Closing meeting 122
Closure 170, 171
Collecting the data 114, 115
Commercialized product
      approach 50, 70, 71
Commitments 185, 186, 196
Communicating with the
      auditee 106, 125, 126
Communication 7, 8, 128
Communication skills (auditor)
      82-84
Competence (auditor) 80, 81
Complaint 117
Complaint files 61
Complaints (product)
      approach 50, 61, 62
Completed records 54-56
Compliance 2, 8, 19, 96, 97
Compliance audit. *See* GMP
      audit
*Compliance Program Guid-
      ance Manual* (CPGM) 47
Components 66
Computer systems 59
Computer systems approach
      50, 69, 70
Conclusions 160
Conduct the audit 15, 16,
      106-124
Confidentiality agreements 176
Confidential materials 84, 150,
      161
Consistent (records) 56
Contacts 91, 95

Contract acceptor 50, 104,
      173, 174
Contract giver 43
Contract manufacture and
      analysis 29
Contractor audit 42, 50,
      104, 173-181
Contractor requirements
      approach 50
Contractors 65, 122, 173, 174
Corrective actions 168, 169, 180
CPGM. *See Compliance
      Program Guidance Manual*
CQA 80
Critical deficiency 96, 147, 149
Cross-contamination 115
Cues, visual 129
Current supplier 42
Current vendor 174

D
Data
   analysis 133-142, 180
   collection 114, 115
   quality and integrity
      approach 50, 62, 63
   raw 117
Deficiencies 96, 97, 119-123,
      137-140, 149
Deployment of resources 7
Desk audit 109-112, 114
Detailed report 144-146
Development report 72
Deviation 117, 186
Deviations approach 50, 56, 57
Direct (records) 56
Disagreements (on facts) 167
Disagreement (with audit
      report) 195
Discussion of findings 119
Dispensing 73
Documentation system
      approach 50, 54-56
Documentation techniques 115
Documenting findings 118
Documents, examples 118
Documents requested from
      auditee 101, 104, 191, 192
Drawbacks of checklists 99
Due diligence audit 45

E
EC. *See* European Community
Education (auditor) 78
Electronic signatures 70
Environmental monitoring 74
Ethics 84, 85
European Community 3
European Community GMP
    Guidelines 29, 30, 79, 178
Evaluation process 110, 111, 166
Example documents 118
Excellence, areas of 122
Expectations 154, 157
Experience (auditor) 78
Experts 189, 190
External audit 8, 39
Eye contact 129

F
Facilities and equipment
    approach 50, 67-69
FDA. *See* United States Food
    and Drug Administration
FDA *Compliance Program
    Guidance Manual*
    (CPGM) 47
FDA Form 483 34, 96, 176, 193
FDA *Investigations Operations
    Manual* (IOM) 34, 47
FDA policy CPG 7151.02 33
FDA Pre-Approval Inspection.
    See Pre-Approval inspection
File documents 91, 104
Filter selection 74
Final report 18
Findings 153
    discussion 119
    documenting 118
    initial 122
First-in/first-out (FIFO) 73
First-party audit 8, 9, 39
Focused audit 43
Follow-up and closure 16,
    17, 163-171, 180, 196
Foreign Inspection Guide 104
Format (report) 145, 155-157
Functional documents 54, 55

G
General GMP operations
    approach 50, 72-76

Gestures 130
Goal 14, 40, 50, 90, 146
GMP audit 2-6
GMP Quality System
    approach 50-53
GMP regulations 47. *See
    also* cGMP
Guidelines 188, 189
*Guidelines for Auditing
    Quality Systems* 78
"Gut" reactions 131

H
High purity water systems 95
Historical documents 91, 104
Hosting auditors 184, 185
Housekeeping 68
HPB. *See* Canadian Health
    Protection Branch

I, J
Identity, product 115
Impact of observations 120
Improvement 7
Informal inspection 11, 41
Information sharing 121
Initial findings 122
Inspection 5, 10, 47
Installation qualification (IQ)
    58, 72
Integrity 84, 193
Internal audit 8, 39, 79
Internal audit report 151-154
Internal cover memo 150
International Organization for
    Standardization. *See* ISO
Interpersonal skills (auditor)
    82-84
Interrogation 126
Interviewing 116, 125-132
Investigational product
    approach 50, 71, 72
*Investigations Operations
    Manual* (IOM) 34, 47
ISO 3, 4, 78, 80
ISO 9000 26, 31, 44, 81
ISO 9001 23, 24, 79, 164,165
Isolated deficiencies 137-140
Isolated occurrence 110, 119

K

Key contacts, auditee 91, 95

L

Labels and labeling 67, 75
Lab notebooks 63, 119
Laboratories approach 63, 64
Laptop computer 119
Lead auditor 86, 91, 113
Legible (records) 55

M

Maintenance 67, 68, 76
Maintenance, equipment 64
Major deficiency 96, 147, 149
Management audit 41, 42
Management support 17
Manufacturing 74
Marketed products 70
Master plan, validation 58
Meetings 112, 114, 122
Minimum requirements 20
Mock inspections 44

N

New vendor 174
Noncompliance 96, 97
Nonprocess-related
    deficiencies 139

O

One-time occurrence 110
Opening meeting 112, 114,
    178, 179
Operation qualification (OQ)
    58, 72
Outcomes 17, 18
Out-of-specification (OOS)
    56, 117
Outstanding compliance 97

P

Packaging and labeling 75
Parenteral manufacturing.
    See Aseptic/parenteral
    manufacturing
Pareto Principle 135
Performance model 140
Performance qualification
    (PQ) 58

Periodic quality reviews 70
Permanent (records) 55
Personal attributes (auditor) 80
Pest control 67, 76
Phases of the audit process
    14-17, 89, 109
Policies 32-37
Potential suppliers 42, 52
Pre-Approval inspection 21,
    186, 188
Pre-Approval inspection
    program 44
Preliminary audit findings 144
Preliminary report 194, 195
Preparation 14, 15, 89-107,
    175-181, 184-190
Preparatory audit 43, 44
Preparatory general
    inspection 50
Preparatory PAI audit 50
Preparing checklists 99, 100
Problems 56
Procedures. See SOPs
Process-related deficiencies 138,
    139
Product 114, 115, 120
Products to audit 92
Prompt (records) 56
Proposed suppliers/contracts
    122
Proprietary information 161
Protocols, validation 58, 104
Purchasing 73
Purity, product 115
Purpose 39, 91

Q

QC labs 50
Qualifications for GMP
    auditors 86, 87
Qualified (auditor) 78, 80
Quality audit 4, 21-25, 41,
    42, 79. See also GMP
    audit
Quality management 29
Quality/Management audit 50
Quality, product 115
Quality system 13
Quality System approach
    50-53

Quality system components
    13-15, 43, 53
Quality System Regulations
    for medical devices 4, 35
Questionnaires 100

R
Raw data 117
Raw materials 66
Raw material supplier certifi-
    cation program 28, 29
Reactions, "gut" 131
"Readily available"
    (documents) 191, 192
Receiving. See Warehousing/
    receiving
Recommendations 154
Records. See Completed records
Regulatory audit 46, 47
Regulatory knowledge (audi-
    tor) 81-84
Reluctant interviewees 130
Report 16, 91, 143-162
Reports, validation 58
Requirements 20
Requirements (auditor) 79
Response (to audit report)
    163-166, 183, 195
Responsibilities of audit team 95
Retrospective validation 59
Retrospective view 5

S
Safety, identity, strength,
    purity, and quality
    (SISPQ) 96, 97, 120
Safety, product 115
Sampling 117, 118
Satisfactory/adequate
    compliance 97
Scheduling an audit 90
Scope of the audit 90, 92, 93
Scope of the observation 119
Second-party audit 9, 39
Selection of team members 96
Self-analysis by team
    members 141
Self-audit. See Self-inspection
Self-confidence 130

Self-evaluation 185
Self-inspection 3, 4, 29, 30,
    41, 79
Sharing information 121
SISPQ See Safety, identity,
    strength, purity, and quality
Site audit 112
SOPs 31-35, 104
Spoken communication 128
Sponsor 8
Stability data 60
Stability testing and monitor-
    ing approach 50, 60, 61
Standard format 145, 158, 159
Standardization 7
Standard operating proce-
    dures. See SOPs
Standards 91, 96-98, 113
Strength, product 115
Success, interview 127
Supplier audit 42, 50, 104
Supplier requirements
    approach 50, 64-67
Suppliers 25, 28, 29, 42, 52,
    122, 174
System conformity audit 112
Systemic deficiencies 137-140
Systemic failure 119
System suitability audit 111

T
Talking with people 116, 125
Team leader 91, 110
Technical knowledge
    (auditor) 81-84
"Themes" 134
Third-party audit 9, 10, 39, 164
Third-party auditor 42
Trace-back technique 116, 117
Tracking deficiencies 138
Training 8, 76, 78, 103, 117,
    140, 167
Trending 70
Trust levels 168
Truthful (records) 56
21 CFR Parts 210 and 211
    20, 21, 79
21 CFR Part 820 21-27, 79

U

United States Food and Drug
    Administration 4, 20-27,
    46, 47
U.S. cGMPs for Finished
    Pharmaceuticals 20, 21
U.S. cGMPs for Medical
    Devices: Quality System
    Regulation 21-27, 79
U.S. Drug cGMPs 79
Utilities 75

V

Validation 58, 59
Validation and change
    control approach 50, 57-59
Validation protocols 69, 104
Validation records 64, 104
Vendor audits 36, 175-181
Vendor requirements
    approach 64-67
Vendors 20, 25, 42, 65, 104,
    173-181
Verification of corrective
    actions 168, 169
Verification of perceptions
    120, 121
Visual cues 129

WXYZ

Warehousing/receiving 73
Wastepaper 115
Water systems, high purity 75
What to expect (auditor) 85
Working papers 33, 91, 106,
    160, 170
Work instructions 33
Wrap-up meeting 122, 144

Printed in the United States
by Baker & Taylor Publisher Services

Printed in the United States
by Baker & Taylor Publisher Services